Synthese Library

Studies in Epistemology, Logic, Methodology, and Philosophy of Science

Volume 417

The aim of *Synthese Library* is to provide a forum for the best current work in the methodology and philosophy of science and in epistemology. A wide variety of different approaches have traditionally been represented in the Library, and every effort is made to maintain this variety, not for its own sake, but because we believe that there are many fruitful and illuminating approaches to the philosophy of science and related disciplines.

Special attention is paid to methodological studies which illustrate the interplay of empirical and philosophical viewpoints and to contributions to the formal (logical, set-theoretical, mathematical, information-theoretical, decision-theoretical, etc.) methodology of empirical sciences. Likewise, the applications of logical methods to epistemology as well as philosophically and methodologically relevant studies in logic are strongly encouraged. The emphasis on logic will be tempered by interest in the psychological, historical, and sociological aspects of science.

Besides monographs *Synthese Library* publishes thematically unified anthologies and edited volumes with a well-defined topical focus inside the aim and scope of the book series. The contributions in the volumes are expected to be focused and structurally organized in accordance with the central theme(s), and should be tied together by an extensive editorial introduction or set of introductions if the volume is divided into parts. An extensive bibliography and index are mandatory.

More information about this series at http://www.springer.com/series/6607

Anne Sophie Meincke
Editor

Dispositionalism

Perspectives from Metaphysics and the Philosophy of Science

Springer

Editor
Anne Sophie Meincke
Department of Philosophy
University of Vienna
Vienna, Austria

Centre for the Study of Life Sciences (Egenis)
University of Exeter
Exeter, UK

Synthese Library
ISBN 978-3-030-28721-4 ISBN 978-3-030-28722-1 (eBook)
https://doi.org/10.1007/978-3-030-28722-1

This Springer imprint is published by the registered company Springer Nature Switzerland AG.
The registered company address is: Gewerbestrasse 11, 6330 Cham, Switzerland

To my son Friedrich

Acknowledgements

This volume is a key outcome of the research project *Powers and the Identity of Agents*, which was funded by the Austrian Science Fund (FWF), project number 23059, and led by Edmund Runggaldier as the Principal Investigator at the Department of Christian Philosophy of the University of Innsbruck, Austria. I was a Research Fellow on this project from 2011 to 2014, and I have continued thinking about powers and dispositions ever since.

The majority of the papers in this volume originated at the conference "The Ontological Commitments of Dispositionalism", funded by the grant and held in Innsbruck, Austria, from 31st July to 2nd August 2013. I am grateful to all participants in that event, speakers and audience, to the Austrian Science Fund, as well as to the Vice-Chancellor for Research and the International Relations Office of the University of Innsbruck for additional financial support. Thanks are also due to Daniel Wehinger, Monika Datterl, and Ksenia Scharr for assisting me in organising the conference.

This volume would not have been possible without the enthusiasm and hard work of its contributors. I am deeply indebted to all of you for your willingness to take part in this project and for your patience along the way. An anonymous reviewer made helpful comments on an earlier version of the book manuscript, which were greatly appreciated. I am also grateful to the project members, Christian Kanzian, Josef Quitterer, Edmund Runggaldier, and Daniel Wehinger, and to everybody else with whom I had the chance to discuss matters concerning powers and dispositions in the past years. Special thanks in this latter regard I would like to offer to Rani Lill Anjum, Alexander Bird, Thomas Buchheim, John Dupré, Florian Fischer, Steven French, Ruth Groff, John Heil, Ludger Jansen, Geert Keil, Anna Marmodoro, Thomas Müller, Stephen Mumford, Timothy O'Connor, John Pemberton, Markus Schrenk, Barbara Vetter, and Neil E. Williams.

There is no book without a publisher. Therefore, I would like to express my gratitude to Otávio Bueno for including this volume in the *Synthese Library* and to Ties Nijssen and Christopher Wilby from Springer for their practical support.

In the course of working on this project, my son Friedrich grew up in the vicinity of my desk from a toddler to a young boy. I dedicate the book to him, who is the most precious treasure of my heart.

Exeter, UK
June 2019

Anne Sophie Meincke

Contents

Contributors

Rani Lill Anjum School of Economics and Business, Norwegian University of Life Sciences (NMBU), Ås, Norway

Alexander Bird Department of Philosophy, King's College London, London, UK

Steven French School of Philosophy, Religion and History of Science, University of Leeds, Leeds, UK

Ruth Groff Department of Political Science, Saint Louis University, St. Louis, MO, USA

John Heil Department of Philosophy, Washington University in St. Louis, St. Louis, MO, USA
Department of Philosophy, University of Durham, Durham, UK

Max Kistler Institut d'Histoire et de Philosophie des Sciences et des Techniques, Université Paris 1 Panthéon-Sorbonne, CNRS, Paris, France

Anna Marmodoro Department of Philosophy, Durham University, Durham, UK
Corpus Christi College, University of Oxford, Oxford, UK

Anne Sophie Meincke Department of Philosophy, University of Vienna, Vienna, Austria
Centre for the Study of Life Sciences (Egenis), University of Exeter, Exeter, UK

Stephen Mumford Department of Philosophy, University of Durham, Durham, UK
Centre for Applied Philosophy of Science, School of Economics and Business, Norwegian University of Life Sciences (NMBU), Ås, Norway

Elina Pechlivanidi Department of History and Philosophy of Science, University of Athens, Athens, Greece

Stathis Psillos Department of History and Philosophy of Science, University of Athens, Athens, Greece

Josef Quitterer Department of Christian Philosophy, University of Innsbruck, Innsbruck, Austria

Matthew Tugby Department of Philosophy, University of Durham, Durham, UK

Neil E. Williams Department of Philosophy, University at Buffalo, Buffalo, NY, USA

Chapter 1
Dispositionalism: Between Metaphysics and the Philosophy of Science

Anne Sophie Meincke

Abstract According to dispositional realism, or dispositionalism, the entities inhabiting our world possess irreducibly dispositional properties – often called 'powers' – by means of which they are sources of change. Dispositionalism has become increasingly popular among metaphysicians in the last three decades as it offers a realist account of causation and provides novel avenues for understanding modality, laws of nature, agency, free will and other key concepts in metaphysics. At the same time, dispositionalism is receiving growing interest among philosophers of science. This reflects the substantial role scientific findings play in arguments for dispositionalism which, as a metaphysics of science, aims to elucidate the very foundations of science. In this introductory chapter, I give an overview of the state of the debate and explain the twofold aim of the present collection of essays which is (i) to explore the ontological commitments of dispositionalism and (ii) to discuss these against the background of latest scientific research, by bringing together perspectives from both metaphysics and the philosophy of science. I finally provide a summary of this intellectual journey.

Keywords Dispositionalism · Powers metaphysics · Ontological commitments · Philosophy of science · Metaphysics of science

---

A. S. Meincke (✉)
Department of Philosophy, University of Vienna, Vienna, Austria

Centre for the Study of Life Sciences (Egenis), University of Exeter, Exeter, UK
e-mail: anne.sophie.meincke@univie.ac.at; a.s.meincke@exeter.ac.uk

A. S. Meincke (ed.), *Dispositionalism*, Synthese Library 417,
https://doi.org/10.1007/978-3-030-28722-1_1

1.1 The Debate on Dispositionalism

Dispositional properties are properties that are manifested under specific conditions. They characterise the entities that possess them,[1] however, with respect to their potential rather than actual behaviour. Ascribing dispositional properties to something thus involves certain expectations as to how this entity will behave. If the wine glass that I just dropped breaks, or if my dog barks at the postman as he delivers a parcel, I would not normally be surprised. Glass breaks if struck, and dogs bark if they get excited. I know that this is how things are, at least in typical cases, and this knowledge is encoded in my ascription of fragility to the wine glass and of the ability to bark to my dog.[2]

Dispositional realism, or dispositionalism, is the philosophical view that the dispositional properties possessed by entities are essentially dispositional, i.e., cannot be reduced to non-dispositional, 'categorical' properties. According to the so-called dispositionalist essentialists, this means that at least some of all the properties of entities are irreducibly dispositional. So-called pandispositionalists even believe that all properties possessed by entities are irreducibly dispositional.[3] Dispositionalism is an ontological thesis rather than just an epistemological one. Surely, classifying entities according to their dispositions helps us orientate ourselves in an ever-changing, dynamic world where entities act and react, ourselves included. Different kinds of things have different kinds of disposition, and it is useful to take this into account when navigating through the causal muddle. The point emphasised by the dispositionalist is that there really *are* such dispositions as fundamental features of reality and, likewise, that there *is* causation brought about by things in virtue of, or by means of, their respective dispositional properties. Many dispositionalists indeed speak of dispositions as (causal) 'powers' of entities to stress their causal efficacious and dynamic character (e.g., Harré and Madden 1975; Shoemaker [1980] 2003; Mumford 1998; Molnar 2003; Marmodoro 2010; Mumford and Anjum 2011; Bird et al. 2012; Heil 2003, 2012; Jacobs 2017).[4]

---

[1]I deliberately speak of 'entities' rather than (as many scholars do) of 'objects' to leave open the possibility that also non-object-like entities, such as processes, events, states of affairs, structures etc., can possess dispositional properties. I explain below (in Sect. 1.2) why this is important.

[2]Whether this, as traditionally had been assumed, implies that dispositions are intrinsic properties has become a matter of controversy. Arguments for the thesis that (at least some) dispositions are extrinsic properties have been offered by McKitrick (2003b, 2018) and Choi (2009). See also the discussion in Yates (2016).

[3]Another variety of dispositionalism is the 'powerful qualities view' according to which every property is both dispositional and categorical at the same time, see Heil (2003, 2012), Martin (1997, 2008) and Ingthorsson (2013). Similarly, neutral monism about dispositions (Mumford 1998, pp. 192–195) holds that all properties are of a single type, with the categorical/dispositional distinction representing two different ways of linguistically referring to non-linguistically real properties.

[4]Some dispositionalists distinguish between 'powers' and 'dispositions'. According to Bird (2007), only sparse, fundamental properties have dispositional essences. These he calls 'powers' in contrast to abundant dispositional properties. A similar distinction is made, e.g., by Williams (2019) and by Kistler (Chap. 11, this volume).

As is well-known, in the context of contemporary analytic philosophy the assumption of a robust reality of causation is at least as controversial as that of an irreducibility of dispositional properties. David Hume famously analysed causation as a constant conjunction between spatio-temporally contiguous events and dismissed powers in nature as one of the many "fictions of ancient philosophy" (Hume [1738] 1974, p. 210). In the world, as Hume sees it, "all events seem entirely loose and separate" (Hume [1748] 1975, p. 74). To wit, the constant conjunction between events that he claims constitutes causation exists in our minds, not in nature.[5] Neo-Humean philosophers have followed Hume, notably David Lewis in defending what he calls *Humean supervenience*, "the doctrine that all there is in the world is a vast mosaic of local matters of particular fact, just one little thing and then another" (Lewis 1986, p. ix). According to Lewis, all facts about the world supervene on the contingent spatio-temporal distribution of local intrinsic qualities. These non-dispositional, categorical properties are what they are regardless of the causal roles into which they may enter. They lack dispositional essences.

Neo-Humean philosophers have made considerable efforts to defend this view, thereby appealing to a scientific worldview. The apparent mysteriousness of dispositional properties that can be possessed by entities in a non-manifest state was aimed to be removed by analysing dispositional ascriptions in terms of counterfactual conditionals much in the same way as was the apparent mysterious causal connection between events (Hume [1748] 1975, e.g., p. 76; Lewis 1973, 1986).[6] However, as the debate on masks or antidotes, finks and mimics shows, the conditional analysis faces difficulties that many believe to be unsurmountable (for a detailed critical discussion see, e.g., McKitrick 2018, ch. 2; for a recent endorsement see Choi 2018). This reinforces the view that the concept of a disposition possesses an ontological foothold in reality. Additionally, it is far from obvious that science supports categoricalism or quidditism as assumed by the neo-Humeans. It has been claimed that "science finds only dispositional properties all the way down" (Blackburn 1990, p. 255; see also Vetter 2015, p. 8 f.).

It therefore does not come as a surprise that recent years have seen a rise of theories that ground causation, both as experienced in everyday life and as described and investigated by the sciences, in the presumably irreducibly real dispositional properties of entities. Opposing "the dead world of mechanism" (Ellis 2002, p. 60) in which all motion is the mechanical motion of intrinsically inert things, dispositionalists maintain that things are intrinsically active by possessing causally efficacious properties which are essentially defined by their causal roles. This is to insist that the events we observe in the world are not as loose and separate as Hume surmised. According to dispositional realism, how things behave flows from their

[5]Remarkably, though, this involves the assumption of "belief-determining dispositions" (Joy 2013, p. 75).

[6]On the parallels between the conditional analyses of causation and dispositions see Handfield (2009b).

dispositional essences rather than being a matter of the contingent relations between contingently distributed local facts.

Some dispositionalists have concluded that the laws of nature, as being determined by the dispositional essences of properties, are not contingent but metaphysically necessary (Ellis 2001; Bird 2007); others argue for a non-Humean ('realist') version of nomological anti-realism (Mumford 2004; similar Cartwright 1989). Dispositional realism has served as a basis for the proposal of a strong 'this-worldly' modality without possible worlds (Vetter 2015), and it has been suggested that dispositions, by tending towards their manifestations rather than necessitating these, bring with them their own modality lying between necessity and contingency, the so-called dispositionality (Mumford and Anjum 2011; Anjum and Mumford 2018). Furthermore, dispositionalism has entered debates in the philosophy of action on agency and free will (Groff and Greco 2013, part III; Hyman 2014; Spann [née Meincke] and Wehinger 2014). For instance, libertarianism has been defended on the basis of an appeal to (agent-) causal powers (e.g., Lowe 2008; O'Connor 2009; Steward 2012; Mumford and Anjum 2014, 2015a, b; Groff 2019), while the so-called New Dispositionalists refer to dispositions to argue for compatibilism (Vihvelin 2004; Smith 2004; Fara 2005; Fara 2008; Vihvelin 2013; for a critique see Clarke 2009). The concept of a disposition has also been used, with a view to corresponding debates in metaphysics, in debates on the moral status of human embryos (Damschen and Schönecker 2003; Labuda and Baňas 2009; Meincke 2015, 2018) as well as in metaethics, epistemology and political philosophy (e.g., Brower 1993; Yalowitz 2000; Robinson 2006; Gundersen 2010; Robinson 2011; Dumsday 2016; Groff and Greco 2013, parts IV and V; Anjum and Mumford 2018, ch. 9).

1.2 The Twofold Aim of the Present Collection

The burgeoning of dispositionalism, evidenced also by a number of recent edited volumes on the topic (Kistler and Gnassounou 2007; Damschen et al. 2009; Handfield 2009a; Marmodoro 2010; Bird et al. 2012; Groff and Greco 2013; Jacobs 2017), is commonly perceived, and by its proponents often presented (see, e.g., Groff and Greco 2013), as being part of a general revival of a broadly Aristotelian approach to metaphysics. Aristotle emphatically embraced the existence of capacities as sources of change in the world. Whatever change happens does so, according to Aristotle, thanks to the interaction of active and passive capacities of substances. Capacities – powers – are at the centre of Aristotle's concept of causation (Aristotle 1984, *Metaphysics* IX). Yet, we should beware of hastily equating dispositionalism and (neo-) Aristotelianism. Different versions of dispositionalism have been put forward, some of which show significant Humean influences (programmatically, e.g., Handfield 2008). By no means all dispositionalists subscribe to Aristotelian substance ontology; dispositionalism has been combined with (variants of) trope theory such that powers, as tropes, constitute one of several fundamental categories (Molnar 2003; Heil 2003). Maybe dispositions don't need any bearers at all

(McKitrick 2003a, 2018; Mumford 2006), or at least some dispositions don't need bearers and we can generally be pluralistic about their nature (McKitrick 2018). It is, hence, anything but clear that there is *the* dispositionalist philosophy, or *the* powers metaphysics, and that this metaphysics is Aristotelian. Instead, the ontological commitments vary from version to version.

One of the aims of the present collection of essays is to explore the ontological commitments of dispositionalism. In accordance with the above considerations, this has to be understood correctly as an investigation into the various possible ways to spell out the ontological profile of dispositionalism; ways both already realised and still to discover. The idea is not to streamline dispositionalism but rather to appreciate its ontological diversity and richness. This includes bringing to the fore the respective motivations leading to specific ontological claims. If one, for instance, believes that dispositions or powers require substances as bearers, why is that? What surrounding ontological assumptions are in place that support this view? And what are the implications? Similar questions apply if one prefers combining realism about dispositions and causation with other kinds of ontology. Interestingly, the reasons, emerging in the course of the discussion, for specific views about what powers are and how they work are not necessarily strictly tied to the commitment to dispositionalism itself. They may derive from views on separate subject matters, such as the questions of how to account for persistence through time or how to understand processes, and more generally they may depend on broader ontological orientations. This puts dispositionalism at the centre of fundamental metaphysical oppositions and battles, and raises the question of how to make, if any, a well-founded choice between its different versions.

An important authority to turn to in the quest for a fair assessment of dispositionalism is science, and this indeed constitutes the second major aim of this volume: to evaluate recent dispositionalist proposals in metaphysics in the light of the latest scientific investigations into the structures of reality. As already indicated, dispositionalists generally take science to support their case. Accordingly, there is a growing tendency to promote dispositionalism as a variant of a 'metaphysics of science', i.e., a metaphysics that explicates the metaphysical preconditions of science, taking seriously its empirical findings (Mumford and Tugby 2013a, b; Schrenk 2017; Dumsday 2019).[7] At the same time, an increasing number of philosophers of science are taking up dispositionalist ideas, aiming to make them fruitful for the understanding and advancing of science. These endeavours can be generic, such as Chakravartty's dispositions-based defence of scientific (semi-) realism (Chakravartty 2007, 2017), or focused on specific sciences. Traditionally, physics, being commonly regarded as the fundamental science, has attracted most of the attention (e.g., Thompson 1988; Hüttemann 2009). More recently, however,

[7]In contrast, Ladyman and Ross (2007) who likewise subscribe to the programme of a metaphysics of science dismiss a commitment to causal powers for allegedly being part of a metaphysics that "aims at domesticating scientific discoveries so as to render them compatible with intuitive or 'folk' pictures of structural composition and causation (p. 1; see also p. 3 f.).

biology, too, has been considered, both as a resource for backing up arguments for dispositionalism as a metaphysical thesis (Mumford and Anjum 2011, ch. 10), and as a field of research that can benefit from making use of the concept of a disposition (Hüttemann and Kaiser 2018; Austin 2017; see also Meincke 2015, 2018). Reference to dispositions and powers is increasing in the philosophy of science,[8] a development that is prompting intra-disciplinary reflection and controversy (Psillos et al. forthcoming). Not everybody agrees that dispositionalism is supported by science. Reasons range from Hume-inspired general reservations against the concept of a disposition over doubts concerning the possibility of bare dispositions to arguments drawing on specific claims about, or interpretations of, pieces of science, mostly physics. Some of these critical voices from philosophers of science are included in the present collection of essays.

The foregoing has made evident, I hope, the unique scope of this collection which (i) elucidates the ontological profile of dispositionalism by exploring its ontological commitments, (ii) confronts the perspective from metaphysics with perspectives from the philosophy of science and (iii) brings together viewpoints of both scholars sympathetic with dispositionalism and sceptics. I take it to be a general truth that the picture of reality with respect to its ultimate structure needs to be drawn as a joint venture of metaphysics and the philosophy of science. This applies all the more to a prospective dispositionalist picture of reality which would substantially rely on science. Accordingly, dispositionalists should take seriously criticism raised by philosophers of science. By fostering a constructive dialogue between metaphysicians and philosophers of science, the book aims to push forward thinking about dispositions and powers in both disciplines. It also aspires to spark debates on aspects so far not at the centre of attention, such as the nature of mental dispositions, the relationship between dispositionalism and process ontology, lessons from statistics for a dispositionalist theory of causation (and vice versa), and organic powers.

1.3 A Journey Guide

The volume contains original contributions by leading and up-and-coming proponents of the debate on dispositionalism. The chapters are ordered as a continuous journey from more metaphysical viewpoints over more direct engagements with the philosophy of science to perspectives from the philosophy of science itself.

The journey starts with **John Heil**'s (Chap. 2) investigation into the overall metaphysical picture of reality in which causal powers find their natural home. This picture, he argues, is a substance-property ontology. Powers are properties of substances, i.e., powerful ways substances are. Substances, according to Heil, are the fundamental entities described by physics; that is, they are simple, though not

---

[8] Thus, e.g., one of the 81 chapters of the *Companion to the Philosophy of Science*, edited by W. H. Newton-Smith, is on "Dispositions and Powers" (Harré 2000).

necessarily thing-like. Causation has to be understood as the internal relation of powers possessed by substances to their manifestations which happen through interaction with mutual manifestation partners. Heil defends his view against the charge of reductionism and explains why it does not support a perdurantist analysis of persistence.

Ruth Groff (Chap. 3), too, thinks that powers need substances as their bearers. She makes this case by addressing the question of what it is that does the doing in powerful causation. Groff argues that it is not powers but powerful things that do the doing, which is to emphasise that powers, like any properties, can only conceptually, not ontologically, be separated from their bearers. Property-less, and, hence, powerless, substrata ('bare particulars') could not do anything, and neither could the abstract properties as which pure powers would have to be regarded. From an anti-passivist view, the powers-as-doers view can also not be backed up with an appeal to trope theory: clusters of powers are still clusters of abstractions and, hence, cannot do anything.

Josef Quitterer (Chap. 4) explores the ontology of mental dispositions, looking at the two common types of account. He shows that, regardless of whether mental dispositions are taken to be mental states that become manifest by being mentally represented or whether they are conceived as mental powers that are manifested by being exercised through action, there is either no explicit assumption of an enduring bearer of mental dispositions or the assumed enduring bearer fails to qualify as an agent because all the causal work is done by mental events or powers. Quitterer argues for an understanding of mental dispositions that strengthens the case for agent causation as opposed to event-causal accounts of action causation.

Anna Marmodoro (Chap. 5) presents aspects of her theory of Power Structuralism, according to which the building blocks of reality are pure powers, i.e., powers that lack a categorical base. Elaborating on the theory's distinctive notion of actuality as activity, Marmodoro argues that the activity of powers can take the form not only of interaction, i.e., change of another, but also appears as non-interactive activity, i.e., self-change. Thanks to this latter type of manifestation of what she calls intransitive powers, the objection that a powers-only ontology, such as Power Structuralism, construes actuality from mere potentiality can be refuted. This objection rests on the mistaken assumption that actuality is inertness.

Neil E. Williams (Chap. 6) asks to which ontological category the manifestations of powers belong and answers that manifestations are best thought of as neither events, nor processes, nor properties, but rather as very short-lived states of affairs. This, he argues, enables us to make sense of cases where – as we usually would describe it – a certain power begins to manifest itself but fails to completely manifest itself due to an antidote interfering. Such protracted manifestation processes have to be understood as supervening upon sequences of short-lived states of affairs such that each state of affairs contains the powers for the subsequent state of affairs. This, Williams claims, avoids the looming contradiction that, in the type of case at issue, a power's manifestation both does and does not occur.

Anne Sophie Meincke (Chap. 7) discusses how dispositionalists ought to account for persistence. She argues that Neil Williams's proposal of powerful perdurance appears to represent a legitimate version of dispositionalism and thus provides good reason to reject Stephen Mumford's claim that dispositionalists ought to be endurantists or, on a weaker reading, that they ought not to be perdurantists. However, it is a different matter which version of powerful persistence one finds convincing. This, it turns out, importantly depends on what one thinks processes are. Rebutting Williams's concept of process, Meincke argues for a process ontological account of powerful persistence.

Rani Lill Anjum and **Stephen Mumford** (Chap. 8) challenge the belief commonly shared by both Humeans and anti-Humeans that regularity is essential to causation. Dispositionalism, they argue, provides a third way between the Humean account of causation as a contingent regularity and the anti-Humean account of causation as a necessary regularity, provided dispositions are understood as tendencies, i.e., as powers that dispose towards their effects in a way stronger than pure contingency but weaker than necessity. Looking at the statistical analysis of large-population data in medicine, Anjum and Mumford argue that the probabilities to be observed are best explained in terms of such tendencies, i.e., in terms of individual propensities, rather than supporting a regularity account of causation.

Elina Pechlivanidi and Stathis Psillos (Chap. 9) take issue with Mumford and Anjum's view that powers dispose towards their manifestations without necessitating these and, i.e., with the assumption of a sui generis dispositional modality ('dispositionality'). In their critique, Pechlivanidi and Psillos focus on what constitutes an essential element of this theory: the thesis that powers can be represented as vectors. This thesis, they argue, is mistaken in that it overlooks crucial differences between the magnitude and directedness of vectors and the alleged magnitude and directedness of powers. Powers are not vectors, nor are they analogous to vectors. Pechlivanidi and Psillos conclude that Mumford and Anjum's case for a dispositional modality fails.

Alexander Bird (Chap. 10) presents a dispositional account of causation which suggests that causes are conditionally sufficient, rather than necessary, for their effects. In contrast with, e.g., Mumford and Anjum's dispositional account of causation, according to Bird's account, the disposition is not the cause but what is stimulated by the cause so as to produce a certain effect. The disposition is a causal condition of the effect. Bird argues that the concept of subjunctive sufficiency, which, though a kind of necessitation, does not require that causation rule out the metaphysical or nomic possibility of interference, is preferable to Mumford and Anjum's concept of a dispositional modality. Bird, remains, however, sceptical as to whether from any dispositional account of causation arguments for a powers metaphysics can be drawn.

Max Kistler (Chap. 11) opposes versions of dispositionalism according to which laws of nature have either only derivative status or do not exist at all in addition to natural properties. Against such views, Kistler maintains that even within a metaphysical framework that assumes the existence of natural properties conceived as powers, laws of nature are required to make sense of science such as, and in the first place, physics, where powers and laws together figure as truth-makers for disposition attributions. Kistler explains that to combine dispositionalism with realism about laws of nature is to endorse a thin notion of a property which does not determine which other properties an object possesses, as opposed to a thick notion which contains all its dependency relations to other properties.

Steven French (Chap. 12), in contrast, raises fundamental doubts as to whether dispositionalism is compatible with physics. In fact, he claims, it is not – in much the same way that most of metaphysics has failed to keep up with recent developments in science. The main reason why, according to French, dispositionalism has to be dismissed is that it cannot accommodate the symmetry principles that function as constraints on the fundamental laws in physics. This difficulty, he further argues, cannot be circumvented by eliminating laws of nature from our metaphysical pantheon. However, following a 'toolbox' approach to metaphysics, French considers the possibility that Barbara Vetter's notion of potentiality may be usefully combined with structural realism.

Matthew Tugby (Chap. 13), finally, explores if dispositionalists can assist philosophers of biology with the task of distinguishing organisms from non-organisms. Assuming that the difference between an organism and a non-organism consists in the kind of power that each has, Tugby, drawing on David Oderberg's work, considers the possibility that all and only organisms possess self-directed ('immanent') powers, whereby the possessor of the power is also the subject of the power's effect. He argues that while this seems to be a plausible approach to accounting for the distinctive teleology of organic powers, it fails to draw a sharp distinction between organic and inorganic powers. Tugby suggests that we accept that the distinction between organic and inorganic powers is a matter of degree.

References

Anjum, Rani L., and Stephen Mumford. 2018. *What Tends To Be: The Philosophy of Dispositional Modality*. London/New York: Routledge.

Aristotle. 1984. *Metaphysics*. In *The Complete Works of Aristotle, Vol. II*, ed. J. Barnes. Princeton: Princeton University Press.

Austin, Christopher J. 2017. Evo-devo: A Science of Dispositions. *European Journal for Philosophy of Science* 7 (2): 373–389.

Bird, Alexander. 2007. *Nature's Metaphysics: Laws and Properties*. Oxford: Oxford University Press.

Bird, Alexander, Brian Ellis, and Howard Sankey, eds. 2012. *Powers, Properties and Structures. Issues in the Metaphysics of Realism*. London: Routledge.

Blackburn, Simon. 1990. Filling in Space. *Analysis* 50 (2): 62–65.

Brower, Bruce W. 1993. Dispositional Ethical Realism. *Ethics* 103 (2): 221–249.

Cartwright, Nancy. 1989. *Nature's Capacities and Their Measurement*. Oxford: Oxford University Press.

Chakravartty, Anjan. 2007. *A Metaphysics for Scientific Realism: Knowing the Unobservable*. Cambridge: Cambridge University Press.

———. 2017. *Scientific Ontology: Integrating Naturalized Metaphysics and Voluntarist Epistemology*. Oxford: Oxford University Press.

Choi, Sungho. 2009. The Conditional Analysis of Dispositions and the Intrinsic Dispositions Thesis. *Philosophy and Phenomenological Research* 78 (3): 563–590.
———. 2018. Dispositions. *The Stanford Encyclopaedia of Philosophy* (Fall 2018 Edition), ed. Edward N. Zalta, https://plato.stanford.edu/archives/.fall2018/entries/dispositions/. Last accessed on 10 May 2019.
Clarke, Randolph. 2009. Dispositions, Abilities to Act, and Free Will: The New Dispositionalism. *Mind* 118 (470): 323–351.
Damschen, Gregor, and Dieter Schönecker. 2003. In dubio pro embrione. Neue Argumente zum moralischen Status menschlicher Embryonen. In *Der Moralische Status menschlicher Embryonen. Pro und contra Spezies-, Kontinuums-, Identitäts- und Potentialitätsargument*, ed. G. Damschen and D. Schönecker, 187–267. Berlin/New York: de Gruyter.
Damschen, Gregor, Robert Schnepf, and Karsten R. Stüber, eds. 2009. *Debating Dispositions. Issues in Metaphysics, Epistemology and Philosophy of Mind*. Berlin: de Gruyter.
Dumsday, Travis. 2016. Dispositionalism and Moral Nonnaturalism. *Journal of Value Inquiry* 50 (1): 97–110.
———. 2019. *Dispositionalism and the Metaphysics of Science*. Cambridge: Cambridge University Press.
Ellis, Brian. 2001. *Scientific Essentialism*. Cambridge: Cambridge University Press.
———. 2002. *The Philosophy of Nature: A Guide to the New Essentialism*. Montreal/Kingston: McGill-Queen's University Press.
Fara, Michael. 2005. Dispositions and Habituals. *Noûs* 39 (1): 43–82.
———. 2008. Masked Abilities and Compatibilism. *Mind* 117 (468): 843–865.
Groff, Ruth. 2019. Sublating the Free Will Problematic: Powers, Agency and Causal Determination. *Synthese* 196 (1), 179–200 (Special Issue*: Real Possibilities, Indeterminism and Free Will*). https://doi.org/10.1007/s11229-016-1124-y.
Groff, Ruth, and John Greco, eds. 2013. *Powers and Capacities in Philosophy. The New Aristotelianism*. New York/London: Routledge.
Gundersen, Lars. 2010. Tracking, Epistemic Dispositions and the Conditional Analysis. *Erkenntnis* 72 (3): 535–364.
Handfield, Toby. 2008. Humean Dispositionalism. *Australasian Journal of Philosophy* 86 (1): 113–126.
———, ed. 2009a. *Dispositions and Causes*. Oxford: Oxford University Press.
———. 2009b. The Metaphysics of Dispositions and Causes. In *Dispositions and Causes*, ed. T. Handfield, 1–30. Oxford: Oxford University Press.
Harré, Rom. 2000. Dispositions and Powers. In *A Companion to the Philosophy of Science*, ed. W. H. Newton-Smith, 97–101. Malden: Wiley Blackwell.
Harré, Rom, and Edward H. Madden. 1975. *Causal Powers: A Theory of Natural Necessity*. Oxford: Blackwell.
Heil, John. 2003. *From an Ontological Point of View*. Oxford: Clarendon Press.
———. 2012. *The Universe as We Find It*. Oxford: Oxford University Press.
Hume, David. [1738] 1974. *A Treatise of Human Nature, Vol. I*, ed. A. D. Lindsay. London/New York: Dent and Dutton.
———. [1748] 1975. *Enquiries Concerning Human Understanding and Concerning the Principles of Morals*, 3rd ed. P. H. Nidditch. Oxford: Clarendon Press.
Hüttemann, Andreas. 2009. Dispositions in Physics. In *Debating Dispositions. Issues in Metaphysics, Epistemology and Philosophy of Mind*, ed. F. Damschen, R. Schnepf, and K.R. Stüber, 223–237. Berlin: de Gruyter.
Hüttemann, Andreas, and Marie I. Kaiser. 2018. Potentiality in Biology. In *Handbook of Potentiality*, ed. K. Engelhard and M. Quante, 401–428. Dordrecht: Springer.
Hyman, John. 2014. Desires, Dispositions and Deviant Causal Chains. *Philosophy* 89 (1): 83–112.
Ingthorsson, Rögnvaldur. 2013. Properties: Qualities, Powers, or Both? *Dialectica* 67 (1): 55–80.
Jacobs, Jonathan D., ed. 2017. *Causal Powers*. Oxford: Oxford University Press.

Joy, Lynn S. 2013. The Ineliminability of Dispositions in Hume's Rejection of Causal Powers. In *Powers and Capacities in Philosophy. The New Aristotelianism*, ed. R. Groff and J. Greco, 69–92. New York/London: Routledge.

Kistler, Max. This volume. Powers, Dispositions and Laws of Nature. In *Dispositionalism. Perspectives from Metaphysics and the Philosophy of Science (Synthese Library* 417*)*, ed. A. S. Meincke, 171–188. Cham: Springer.

Kistler, Max, and Bruno Gnassounou, eds. 2007. *Dispositions and Causal Powers*. Aldershot: Ashgate.

Labuda, Pavol, and Jan Baňas. 2009. Conceptual Analysis of the Potentiality Argument in Favor of Human Embryo's Right to Life. *International Journal of Philosophy*: 49–64.

Ladyman, James, and Don Ross. 2007. *Every Thing Must Go: Metaphysics Naturalized*. Oxford: Oxford University Press.

Lewis, David K. 1973. *Counterfactuals*. Oxford: Blackwell.

———. 1986. *Philosophical Papers, Vol. II*. New York: Oxford University Press.

Lowe, E. Jonathan. 2008. *Personal Agency. The Metaphysics of Mind and Action*. Oxford: Oxford University Press.

Marmodoro, Anna, ed. 2010. *The Metaphysics of Powers. Their Grounding and Their Manifestations*. London: Routledge.

Martin, Charles B. 1997. On the Need for Properties: The Road to Pythagoreanism and Back. *Synthese* 112: 193–231.

———. 2008. *The Mind in Nature*. Oxford: Oxford University Press.

McKitrick, Jennifer. 2003a. The Bare Metaphysical Possibility of Bare Dispositions. *Philosophy and Phenomenological Research* 66 (2): 349–369.

———. 2003b. A Case for Extrinsic Dispositions. *Australasian Journal of Philosophy* 81 (2): 155–174.

———. 2018. *Dispositional Pluralism*. Oxford: Oxford University Press.

Meincke, Anne Sophie. 2015. Potentialität und Disposition in der Diskussion über den Status des menschlichen Embryos: Zur Ontologie des Potentialitätsarguments. *Philosophisches Jahrbuch* 122 (2): 271–303.

———. 2018. Haben menschliche Embryonen eine Disposition zur Personalität? In *Der manipulierbare Embryo. Potentialitäts- und Speziesargumente auf dem Prüfstand*, ed. M. Rothhaar, M. Hähnel, and R. Kipke, 147–171. Münster: Mentis.

Molnar, George. 2003. In *Powers. A Study in Metaphysics*, ed. S. Mumford. Oxford: Oxford University Press.

Mumford, Stephen. 1998. *Dispositions*. Oxford: Oxford University Press.

———. 2004. *Laws of Nature*. London/New York: Routledge.

———. 2006. The Ungrounded Argument. *Synthese* 149 (3): 471–489.

Mumford, Stephen, and Rani L. Anjum. 2011. *Getting Causes from Powers*. Oxford: Oxford University Press.

———. 2014. A New Argument Against Compatibilism. *Analysis* 74 (1): 20–25.

———. 2015a. Freedom and Control: On the Modality of Free Will. *American Philosophical Quarterly* 52 (1): 1–12.

———. 2015b. Powers, Non-Consent and Freedom. *Philosophy and Phenomenological Research* 91 (1): 136–152.

Mumford, Stephen, and Matthew Tugby, eds. 2013a. *Metaphysics and Science*. Oxford: Oxford University Press.

———. 2013b. What is the Metaphysics of Science? In *Metaphysics and Science*, ed. S. Mumford and M. Tugby, 3–28. Oxford: Oxford University Press.

O'Connor, Timothy. 2009. Agent-Causal Power. In *Dispositions and Causes*, ed. T. Handfield, 189–214. Oxford: Clarendon Press.

Psillos, Stathis, Benjamin Hill, and Henrik Lagerlund, eds. Forthcoming. *Causal Powers in Science: Blending Historical and Conceptual Perspectives*. Oxford: Oxford University Press.

Robinson, Luke. 2006. Moral Holism, Moral Generalism, and Moral Dispositionalism. *Mind* 115 (458): 331–360.

———. 2011. Moral Principles as Moral Dispositions. *Philosophical Studies* 156 (2): 289–309.

Schrenk, Markus. 2017. *Metaphysics of Science. A Systematic and Historical Introduction*. London: Routledge.

Shoemaker, Sydney. [1980] 2003. Causality and Properties. Repr. In *Identity, Cause and Mind*, expanded edition. Oxford: Oxford University Press, 206–233.

Smith, Michael. 2004. Rational Capacities. In *Ethics and the A Priori*, 114–135. Cambridge: Cambridge University Press.

Spann [née Meincke], Anne Sophie and Daniel Wehinger, eds. 2014. *Vermögen und Handlung. Der dispositionale Realismus und unser Selbstverständnis als Handelnde*. Münster: Mentis.

Steward, Helen. 2012. *A Metaphysics for Freedom*. Oxford: Oxford University Press.

Thompson, Ian J. 1988. Real Dispositions in the Physical World. *British Journal for the Philosophy of Science* 39 (1): 67–97.

Vetter, Barbara. 2015. *Potentiality. From Dispositions to Modality*. Oxford: Oxford University Press.

Vihvelin, Kadri. 2004. Free Will Demystified: A Dispositional Account. *Philosophical Topics* 32 (1/2): 427–450.

———. 2013. *Causes, Laws, and Free Will. Why Determinism Doesn't Matter*. Oxford: Oxford University Press.

Williams, Neil E. 2019. *The Powers Metaphysic*. Oxford: Oxford University Press.

Yalowitz, Steven. 2000. A Dispositionalist Account of Self-Knowledge. *Philosophy and Phenomenological Research* 61 (2): 249–278.

Yates, David. 2016. Is Powerful Causation an Internal Relation? In *The Metaphysics of Relations*, ed. A. Marmodoro and D. Yates, 138–156. Oxford: Oxford University Press.

Chapter 2
Ontology of Powers

John Heil

Abstract An account of powers situated within a substance–property ontology is sketched according to which substance and property are correlative notions: substances are various ways, properties are ways substances are; every substance is some way or other, every way is a way some substance is. What substances do or would do depends on their properties and properties of substances with which they interact. Properties, on such a view, are powers, powerful ways substances are. Properties are not purely powers, however: properties are powerful qualities. Properties are qualities the identity of which depends on how they would manifest themselves with particular kinds of reciprocal partner. A conception of this kind provides sufficient resources to make sense of modality and causation in a decidedly non-Humean manner. Powers can serve as truthmakers for various modal truths. Causation, understood as the mutual manifestation of powers, is revealed to be a kind of internal relation. With this ontological picture in play, it is possible to provide systematic answers to questions most often posed concerning the nature of powers.

Keywords Power · Disposition · Manifestation · Cause · Law · Truthmaking

> *Remember then: there is only one time that is important—Now! It is the most important time because it is the only time when we have any power.*
>
> *(Tolstoy, 'Three Questions')* (Roy Sorensen called my attention to this passage)

J. Heil (✉)
Department of Philosophy, Washington University in St. Louis, St. Louis, MO, USA

Department of Philosophy, University of Durham, Durham, UK
e-mail: jh@wustl.edu; john.f.heil@durham.ac.uk

A. S. Meincke (ed.), *Dispositionalism*, Synthese Library 417,
https://doi.org/10.1007/978-3-030-28722-1_2

2.1 Ontology

One of the stated aims of this volume is to examine the ontological implications of powers. By way of facilitating discussion, our editor posed four questions:

1. How should powers be ontologically categorised? Do they need bearers? If so, to which ontological category might such bearers belong?
2. Does dispositional realism favour an ontology of substances or events or tropes etc.?
3. Which account of persistence should dispositionalists endorse?
4. Are there irreducible powers only in the microphysical realm or also in the macrophysical realm?

Before indicating how I think these questions ought to be answered, I propose to reflect briefly on the character of ontology generally.

Owing to the ever more technical nature of philosophical discourse and a steady rise of ‘professionalism’, the philosophical world has become increasingly fractured. The onset of blinkered specialisation has fuelled a tendency to see ontological theses as discrete modular affairs capable of being combined, and recombined, mixed and matched so as to achieve some desired result. There are, for instance, various approaches to causation, to powers, to properties, to modality. It is all too common to find philosophers embracing a view of one of these to solve some local problem, then picking and choosing from among the rest so as to fend off particular threats to the favoured position.

I suspect that this practice is abetted by the practice of treating metaphysics as a branch of the philosophy of language. If you begin with language, you lose the unifying force of the universe as we find it. Many philosophers would regard this comment as hopelessly crude. If the idea that we could read off ontology from language is a bad one, the thought that we can derive ontology from the universe as we find it is surely worse. Where would you begin? Would you start with a list of objects in your vicinity and move outwards? Not only would such a project be hopeless, it would defeat the kind of generality and unification we aim at in ontology.

Language is an indispensable vehicle for our thoughts about the universe, including thoughts bearing on ontological matters.[1] But it is the universe, or ways the universe is, that makes those thoughts true. Knowing what we ‘quantify over’ is to know what truths we accept. But this is not the same as knowing the nature of the truthmakers for those truths. My belief is that fundamental physics gives us our best account of the truthmakers for all the truths that have truthmakers. This is what makes fundamental physics fundamental. Ontology gives us our best systematic account of the elementary categories of truthmaker.

Such a conception is in no way reductionist. I do not claim that all the truths are reducible to—analysable into or derivable from—truths of fundamental physics.

[1] In fact, I have some doubts as to whether this is so; see Heil (2012), ch. 12.

Think about it. It can be true that your cursor moves across your computer screen even though the truthmaker for this truth is an immobile, stationary array of pixels exhibiting a dynamic pattern of illumination. It could be true that a baseball sails over the centre field fence even if the motion of baseballs were at bottom something like the motion of the cursor. Truthmakers for claims about moving baseballs could be wavelike disturbances in fields. To conclude from this that there are no baseballs or that baseballs do not really sail over fences, would indeed be reductionist. But such a conclusion would be deeply misguided.

2.2 The Basics: Substance and Property

Philosophers embarking on ontological projects like to invoke Plato's image of 'carving the beast of reality at the joints'. But reality has myriad joints. The problem is not finding joints, but understanding how categories we deploy in apprehending the universe select the joints they do. I believe that, when you consider these categories at the most general level, you arrive at an ontology of substance and property. Substance and property are correlative categories. Substances are property bearers, properties are borne by substances, properties are particular ways substances are. Any property is a way some substance is, and every substance is some way or other. As I shall note presently, I hold, as well, that substances must be simple: a substance cannot have parts that are themselves substances.

Although this is a very big deal, it is not something I shall attempt to defend here. In discussing powers, however, you need to start somewhere. I leave it to those who favour a different ontology to adjust my remarks to suit their tastes. The important point is just that it would be a mistake to tackle questions as to the nature of powers or dispositions—I use the terms interchangeably—in isolation.

Suppose then that you have substances and properties, with properties being particular ways substances are. A fully particular universe, a universe lacking in general or universal entities, provides all the resources we need for making true the general and universal truths that have truthmakers. This in fact is a historically prominent position, although you would never imagine that this might be so if you confined yourself to mainstream analytic metaphysics. I am tempted to say more here, but I will forebear further discussion, once again counting on you to translate my remarks to suit yourself.

Given properties and substances, where do we locate powers? What is the 'ground' or 'basis' of dispositionality? Substances are bearers of powers. An electron has the power to repel other electrons and to attract positrons. An electron has this power in virtue of being negatively charged. An electron is not merely negatively charged, however: an electron has a *determinate* negative charge. And this would seem to be a *property* of the electron, a property the electron possesses along with various others: a particular determinate mass, for instance, and a particular determinate spin.

So I identify powers with properties and wed properties to substances. To some, this is needlessly profligate. Why substances? Why not simply powers? An ontology consisting of powers alone (or powers standing in various relations to one another) might be valued for its simplicity. Science seems to demand powers, but substances appear to be distinctively philosophical posits, mysterious substrata, somethings we know not what, artefacts perhaps of provincial subject–predicate forms of speech.

Although philosophers like to wield Ockham's Razor as a kind of sword of virtue, it is self-defeating to invoke parsimony as a tool in theory construction. Simplicity comes into play only in the end game, only in comparisons among fully developed theories with a common subject matter. One theory can be simpler than another in one respect, but only at the cost of greater complexity elsewhere. Whether an ontology of powers alone (or powers plus relations) is simpler, all things considered, than an ontology of substances and properties, is by no means obvious.

But why suppose we need substances? I described substances and properties as correlative: if you have one, you have the other. Philosophers who regard themselves as savvy about science sometimes evince scepticism about substances on the grounds that the universe as described by fundamental physics looks anything but substantial. Perhaps our universe is not a universe of 'things' at all, perhaps it all comes down to fields pervading space, dynamically evolving over time, or the wave function unfolding in an *n*-dimensional configuration space.

The thought that a substance–property ontology is committed to a kind of corpuscular conception of 'things' is premature however. The ontology leaves open the nature of substances. The substances might be corpuscular bits of matter, but they might be fields, or space-time itself, or multidimensional configuration spaces, or something as yet unfathomed. Particles are candidate substances, but particles might be virtual, particles might be disturbances in fields, dynamic wave-like eddies in the fundamental stuff. In that case what you might have thought were substances—electrons, say—will turn out to be modes of some substance, not substances themselves but ways some substance is. The same would be true of events and processes. If there are such things, they would seem to involve substances and not replace them.

A commitment to a substance–property ontology is a commitment to the idea that, whatever the final theory, there will be items—or an item—playing the substance role, somethings that are various ways, and ways those somethings are. I am not aware of any scientific theory that fails to fit this model, although there are certainly philosophical interpretations of scientific theories that deny it.

2.3 Powers as Relations

On one of these interpretations the most likely ontology of future physics is an ontology of pure relations. The guiding idea is that science is in the business of identifying powers and powers are ultimately to be understood wholly in terms of relations: a power is what it is owing to relations it bears to all the other powers. This way of putting it is somewhat misleading, however, for, on the view under consideration, there is no *its* standing in relations to other *its*. Powers lack 'intrinsic natures'. Powers are constructed on or from relations. The result is a species of Humean supervenience according to which everything supervenes on a vast network of relations.

For most of us outsiders it would be natural to think that relations require nonrelational *relata*. To have a relation you must have somethings to relate. The relational view is more ambitious. Relations are taken to be prior to *relata*: *relata* are what you get when you have the relations.

If such a conception has any plausibility at all, it does so only by construing powers as relations. If you thought that powers were relations, and you thought that powers are all there is, you would have a purely relational world. But why think powers are relations?

You might start with the idea that powers are individuated by what they are powers *for*. Water has the power to dissolve salt, salt the complimentary power to be dissolved by water. A case can be made for pairing powers up this way. (Think of the long Aristotelian tradition of 'active' and 'passive' powers or Sydney Shoemaker's forward- and backward-looking powers; Shoemaker 1980, 1998.) So you have powers individuated by powers. Couple this to the idea that the universe as revealed by the sciences is a universe of powers, and it is but a short leap—I exaggerate—to the thought that everything in the universe is constituted by relations. (I leave aside the further question of what the universe *itself* is.)

Although it seems indisputable that powers are individuated by what they are powers for and that powers are typically reciprocal, I do not think that this in any way shows either that powers *are* relations or that properties in virtue of which substances have powers are *purely* powers.[2] I shall address this second point in due course. For the moment I want to focus on relations.

I myself am sceptical that a fundamental ontology is going to include relations. To be sure there are irreducible relational truths, but truthmakers for these truths could turn out to be nonrelational ways the universe is. I believe that this is the best way to understand so-called internal relations. Suppose it is true that Simmias is taller than Socrates. The truthmaker for this truth is Simmias's being a particular determinate height and Socrates's being a particular determinate height. If God wants to make it the case that Simmias is taller than Socrates, God must make Simmias and Socrates and give them each a particular height. (I am assuming that having a particular height is a nonrelational feature of whatever has the height.) The idea that relations could be fundamental in the sense that relata are dependent for their respective identities on what relates them strikes me as having things exactly backwards.

Whether you are with me on this point, I hope you will agree that it is a mistake to construe powers themselves as relations. We *describe* powers relationally by saying

[2]For a different view on these matters, involving the thesis that powers are not relations, but are *relational*, see Marmodoro (Chap. 5, this volume).

what they are powers for, but the powers in no sense depend on their manifestations. An electron does not lose the power to repel electrons if it moves outside the light cone of other electrons. Powers can exist unmanifested. Indeed, a power might be unmanifestable—unmanifestable because the reciprocal power required for its manifestation is spatially or temporally out of reach or simply does not exist. Powers are 'ready to go', but might never go. If you don't like this, you won't like powers.

2.4 Real Powers

Few would deny that the identity of a power is determined by what the power is a power *for*. But a power can be for *different* manifestations with *different* reciprocal manifestation partners.[3] If an electron's negative charge is a power, it is a power to repel other negatively charged particles and a power to attract positively charged particles. Ryle distinguishes single- and multi-tracked dispositions (Ryle 1949, p. 43 f.). If you are serious about dispositions or powers, you will recognise that *most* powers are multi-tracked. The sphericality of a ball is responsible for the ball's rolling—rather than sliding or tumbling—down an inclined plane, for the ball's making a circular—rather than square—concave impression in the carpet, for the ball's displacing a spherical region of the atmosphere, for the ball's looking spherical, and for its feeling spherical.

The key here is reciprocity. Attempts to classify powers by reference to 'stimulus conditions' and manifestations presume a kind of asymmetry absent in nature. Salt is disposed—has a power to—dissolve in water. When you stir a spoonful of salt into a beaker of water what happens? Is it that the water 'triggers' the salt's disposition to dissolve with the result that it dissolves? The description is unperspicuous and highly perspectival. When you stir a spoonful of salt into a beaker of water, the salt and water *interact* to yield a saline solution (see Ingthorsson 2002; Huemer and Kovitz 2003; Heil 2012, ch. 6). This interaction depends on cooperation among powers of the salt and powers of the water. The interaction is continuous and fully symmetrical. There is a sequence here: the salt is introduced into the water, the water and salt interact, and the outcome is a saline solution. But if your interest lies in the nature of causation, you will want to focus on the mutual manifesting of the pertinent dispositions.

The example illustrates both the reciprocity of powers and the reciprocity of causation. Philosophers sometimes appeal to causation to explicate powers, a practice encouraged by reliance on a stimulus–disposition–manifestation model.

---

[3]The phrase 'reciprocal manifestation partners' was a favourite of C. B. Martin's. See Martin (2008).

$$S \rightarrow D \rightarrow M$$

Here, S is a stimulus that causes a disposition, D, to produce a manifestation, M. If you take powers seriously, however, if you take powers on their own terms, you will think that this has the order of explanation backwards. Think of the causal nexus as the mutual manifesting of reciprocal dispositions, and think of the universe as comprising an evolving network of such manifestings. Every manifesting contributes to subsequent manifestings.

2.5 Causation

Are these manifestings law-governed? This depends on what it means to speak of laws as governing. The idea that scientific explanation is a matter of identifying laws presiding over local occurrences is of relatively recent vintage. At one time objects were understood to do what they do because they are as they are. Were you to work out the nature of an object, you would discover what it could do. A conception of this kind locates powers in the objects: objects are *self*-governing. In part because it posed a threat to the omnipotence of God, this kind of Aristotelian view gradually fell out of favour. Powers were stripped from the objects and handed over to God. Understanding why objects do what they do, then, required working out principles on which God determines objects to act, the natural laws.

Nowadays few philosophers who invoke laws in accounts of causation associate laws with God. The laws are *natural*. But if you remove God from the picture, what are laws and in what sense do laws govern? Hume saw clearly that if you stripped the powers from objects and placed them in God, then subtracted God, you lost any sense in which objects could be said to be law-governed. Hume's laws are universal generalisations. This would be one way to go. If you relocate the powers in God, then later give up on God, however, it would seem natural to return the powers to the objects: objects do what they do because they are as they are.[4]

On such a conception what would laws be? Laws might be equations, formulae, principles. These hold, if they do, owing to the powers, owing to the dispositional makeup of the universe. Think of the fundamental laws as distilling the contribution particular kinds of power make to objects possessing them. Thought of in this way, laws need not entail true generalisations. I find this an attractive view in part because it reflects the way scientists often seem to think of laws. Laws are not players. Laws can be true or false, precise or merely approximate, some laws are derivable from other laws.

Suppose something like this is right, suppose the truthmakers for causal claims are the manifestings of powers. Do causes necessitate their effects, or do causes

[4]Another option would be to suppose, as David Armstrong supposes, that laws are general or universal entities, items that bind together other general entities. See Armstrong (1978, 1989, 1997).

simply make certain kinds of effect more probable? Consider a case in which *A*'s cause *B*'s, but not always. Sometimes when an *A* occurs, a *B* occurs, but at other times *B* fails to follow the occurrence of an *A*. Suppose that now an *A* occurs accompanied by a *B*. In this case, in what sense could it be true that the *A brought about* the *B*?[5] In cases in which *A*'s cause *B*'s only sometimes, it is natural to think that *A*'s cause *B*'s only in concert with *C*'s. In that case, *A* and *C* together cause *B*. If *A* occurs in the absence of *C*, *B* does not occur.

Why is this a natural thought? Because it is otherwise hard to understand what is supposed to happen when an *A* brings about a *B* in *this* case but not in others. This has nothing to do with 'exceptionless laws'. It has to do with what is required for something to *bring about* or *produce* something. Here is *A*, and *A* brings about *B*. But there is nothing about *A* in virtue of which *A* does so, no difference between *A*'s that bring about *B*'s and *A*'s that don't. Baffling!

Am I ignoring the indeterministic, probabilistic nature of fundamental physics? No. The indeterministic nature of certain fundamental processes—the decay of a radium atom, for instance—is a reflection, not of a kind of probabilistic causation, but of *spontaneity*. When a radium atom decays, it is not caused to decay in a probabilistic way. The atom decays spontaneously, the decay is *uncaused*. Subsequent *effects* of an atom's decaying are fully deterministic. Once spontaneous unmoved movers are on the scene, indeterminacies proliferate. This is not due to the presence of a special kind of indeterministic or probabilistic causation, however.

So suppose I am right in thinking that causes fully determine their effects. How does this fit with my earlier suggestion that causation is best understood as the manifesting of dispositions? Return to the salt's dissolving in a beaker of water. I described the process as continuous and symmetrical: the salt and the water interact. The dissolving requires both partners (and undoubtedly much else besides) working together cooperatively.

Hold this thought in mind and return to the idea that powers are individuated by their manifestations. More accurately a power is a power to manifest itself in particular ways with particular sorts of manifestation partner. This is its nature, this is what the power *is*. Now return to the dissolving salt. It is of the nature of the salt and the water (and whatever other reciprocal powers are in play) to yield a dissolving. That is what the salt, for instance, *is*: a dissolver-with-these-kinds-of-partner. The identities of the pertinent powers ensure this result: if you have the reciprocal powers in place, you have their manifestations. And this makes it look as though causal relations are in fact a species of *internal* relation: if you have *these* relata, you have *this* relation, this causing.

I do not think it is useful to regard powers as *necessitating* particular kinds of manifestation. One reason this is so is that you can have the powers but, owing to the presence of 'inhibitors', the absence of a particular kind of manifestation.[6] What is

[5]Some readers will resist the idea that causing is a bringing about, but hear me out.

[6]Mumford and Anjum (2011) defend the idea that dispositions operate in accord with a distinctive 'dispositional modality' owing to the possibility of blocking or inhibiting the manifestation of any

determined, what is guaranteed or necessitated depends on a complete ensemble of dispositions. Further, it is too easy to think of necessitation as asymmetrical. Manifestations are determined by reciprocal manifestation partners working together. Manifestations need not follow on the heels of the partnering of powers. Nevertheless, given the powers on hand, and given that a power's identity depends on how that power would manifest itself in concert with particular kinds of reciprocal partner, if you have the powers, you have the manifestations. The relation between powers and their manifestations is internal (see Heil 2012, chs. 6 and 7).

Consider two playing cards leaning against one another and remaining upright on the table. This manifestation is continuous, symmetrical, and contemporaneous with the powers (of the cards, the table, the gravitational field, and whatnot) manifesting themselves as they do. I happen to think (as Kant thought) that the kind of causing illustrated by the playing cards is by far the most common kind. Indeed, I suspect that if you look closely at familiar billiard ball cases of causing, you will find that at bottom their interaction resembles that of the playing cards.

I do not think that philosophers who have embraced, even tentatively, an ontology of powers have fully appreciated the implications of accepting that powers are individuated by their manifestations. To my mind, these implications, far from being a liability, provide a key to understanding causal relations. This is how it should be in ontology. A viable ontology is an organic whole, not a Rube Goldberg contraption that achieves a certain success by bolting together components ad hoc so as to achieve a particular goal.

2.6 Powerful Qualities

I have made a big deal about the identity of powers being tied to their possible manifestations. A substance's possessing a power is its possessing a particular property. But how are we to think of *properties*? Are all properties powers? And when you describe an entity by saying what it does or would do, have you said all you could say about the intrinsic nature of that entity?

I have discussed these questions extensively elsewhere (see, e.g., Heil 2003, 2012). Here I will merely touch on the highlights. Properties are *powerful qualities*. The dispositional–qualitative distinction is one of conception only, what the scholastics and early moderns would have called a distinction of reasoned reason, not a real distinction. Properties do not have two parts, two aspects, two facets. Humeans think that it is easy to imagine cases in which qualities and dispositions vary independently. It is easy to imagine a billiard ball that does not roll down an inclined plane or one that bounces rather than rolls across the table. But billiard balls are

disposition. I regard talk of 'inhibitors', 'antidotes', 'blockers', and 'finks' as perspectival. The presence of an inhibitor, in concert with whatever powers are on hand, yields, not no manifestation, but a different *kind* of manifestation, perhaps the preservation of the status quo.

complex objects with many powers directed toward many kinds of manifestation. Thus, you might fashion a billiard ball out of steel and paint it red. When placed on a magnetic incline, the ball does not roll. A billiard ball made of rubber might bounce when struck, a billiard ball packed with nitroglycerin might explode.

Consider just the ball's sphericality, however. You can regard the ball's sphericality as a quality of the ball. But it is in virtue of being spherical that the ball would roll, would leave a circular concave impression in the carpet, would look spherical, would feel spherical. If you think it is easy to prize apart qualities and powers, try to imagine a spherical object that did not roll, but tumbled, down an inclined plane, that made a square impression in the carpet, that displaced a cubical region of the atmosphere, that looked cubical and felt cubical.

I am not alone in thinking of properties as powerful qualities, but I suspect that what I have to say next will be greeted less enthusiastically. Substances are simple. Given that properties are particular ways substances are, properties are properties of simple substances. Simple substances could be extended in space (hence have spatial parts). Simple substances could be particles, or fields, or space-time, or the universe as a whole. What the simple substances are is an empirical question that belongs to the domain of fundamental physics.

Why think substances must be simple? Start with the thought that substances are fundamental entities, entities that could exist in separation from any other independent entity. A complex made up of substances would be dependent on its parts, the substances that make it up: wholes depend on their parts. Now reflect on substances as property-bearers. Properties are ways substances are. What would it be for a complex made up of property-bearing substances to possess a property? Well, if properties are ways, the properties of such a complex would be ways it is. But ways it is (apart from ways its parts are) would be ways in which its parts are organised. These are not properties of a complex substance, but relations among substances.

I've given you the picture, but little by way of argument. For that, you must go elsewhere. I can however address one sort of worry that many of you are bound to have about any such view. What is left of ordinary objects, complexes made up of substances? These will include billiard balls, trees, human bodies, planets, galaxies, the universe as a whole. Are these unreal? Do we have just the atoms and the void?

Not by my lights. The atoms duly arranged in the void (so to speak), duly interconnected and interacting with one another and perhaps with much else besides, serve as truthmakers for claims about ordinary objects. Such claims can be true, literally, unapologetically true. 'There is a tree in the quad', if true, is made true by (say) a complex, temporally extended arrangement of interacting particles of particular sorts, each with its own particular history. There is, however, no prospect of analysing talk of trees into talk of arrangements of particles, no prospect of expressing application conditions for 'tree' in particle terms, no prospect of purging mention of trees from scientific or everyday discourse, no prospect of describing the universe as we find it without 'quantifying over' trees. This shows that there are endless truths about trees, but certainly not that trees are substances or that predicates true of trees designate properties (if properties are ways substances are).

So it is true to say that there are billiard balls, trees, and the rest, talk of such things is talk of what might turn out to be dynamic clouds of interrelated, interacting particles. (I emphasise that this is only one possibility. The fundamental substances might be fields, or space, or space-time.) What we do not have are the particles with their properties duly interactively organised *plus* billiard balls, trees, and the like with their distinct properties. Think of ordinary objects as what Keith Campbell (1990, pp. 151–155) calls *quasi*-substances, and characteristics we ascribe to them as *quasi*-properties.

I deny that this is a species of nihilism or reductionism. Reduction is a relation among concepts, or predicates, or theories. I expressly deny that reduction of this kind, analytic reduction, is in the cards. Is the view *ontologically* reductive? I am unsure what that would be. Can you reduce one thing, or one property of a thing, to another? Only if you had only one thing to begin with, one thing perhaps could be described or thought of in different ways. But this is not ontological reduction, it is simply the view that one and the same universe contains enough diversity and division to be describable in many, many different nonequivalent ways.

2.7 Substantial Persistence

You are probably thinking that all this takes us far beyond the topic of dispositions. I have gone off on a tangent that threatens to lead us into the Slough of Despond, or, if not there, at least into the swamps of traditional metaphysics from which no one returns entirely intact. My contention, however, is that the ontology of dispositions is not neatly separable from the ontology of everything else. Once you get on the bus you must ride it to the end of the line. I would like to think that I have said enough to demonstrate connections between dispositionality and all the rest, but I have at best gestured in the direction of such connections. I believe they are there and that they are inevitable, but I can't expect you to follow me blindly.

Before concluding, I propose to address—briefly—one more topic. Pretend for a moment that the picture I have sketched has some merit. The universe—duly interrelated, interacting propertied substances—comprises an evolving network of manifestings. One question you might have is whether objects persist over time because their earlier temporal stages are causally responsible for the existence of their later stages, perhaps because these later stages are manifestations of their earlier stages.[7] I believe that a view of this kind is deeply unappealing.

Consider a human being, you for instance. Is the you of right now the product of the you of a moment ago? Think of what is required to hold you together over time. Your continuing to exist requires the massive cooperation of goings on inside and

[7]Neil E. Williams has proposed a dispositionalist account of persistence along these lines, comprising a specific understanding of manifestations of powers as outlined in Williams (Chap. 6, this volume). For a detailed criticism of this view, see Meincke (Chap. 7, this volume).

outside your body. Inside your body vast networks of interconnected cells work continuously simply to keep you intact and alive. Meanwhile, complex forces, including gravitational and electromagnetic forces, pressure from the atmosphere and the like hold you together from one moment to the next. The idea that a 'temporal part' of you, you at t_1, are in some way causally responsible for the you of t_2 lacks credibility.

You might think that all this applies only to complex objects. What of a lone electron moving through space outside the range of potential dispositional partners? Does the electron's persisting under these circumstances require an explanation? If you think, as Descartes thought, that the universe must be recreated (or sustained) at every instant by God, then you would need an account of the electron's persistence. But why think as Descartes did? Descartes had his reasons; what are yours? What of the secular counterpart thesis that the universe comprises distinct and wholly distinct self-contained temporal parts or stages? Again, the motivation for such a view is elusive.

This point is quite independent of the question whether objects have temporal parts. If objects have temporal parts, these would be analogous to spatial parts. Neither kind of part is a substantial part. A complex object has substantial parts duly organised. The object is made up of and dependent on those parts and their interactions. If the object has spatial parts—a right half and a left half, for instance—it is neither made up of nor dependent on those parts; indeed the order of dependence is reversed: *they* depend on *it*, not it on them. Similarly for temporal parts. If you thought of an electron as having temporal parts, this does not turn the electron into a collection of momentary electrons strung together like pearls in a necklace. Temporal parts would be parts of a persisting substance, not temporally instantaneous substances that could interact and add up to a persisting complex substance.

2.8 Dénouement

Let me now return to the four questions posed at the outset and review answers I have proposed for each.

(1) *How should powers be ontologically categorised? Do they need bearers? If so, to which ontological category might such bearers belong?*

Powers are properties of substances. Properties are ways substances are.

(2) *Does dispositional realism favour an ontology of substances or events or tropes &c.?*

The question suggests something itself questionable, namely that you should start with powers, then cobble together an ontology congenial to powers. It could turn out that powers are 'compatible' with lots of candidate ontologies. On my view, however, ontology is, as C. B. Martin liked to put it, a package deal. If powers are legitimate, they should fall out of a plausible ontology.

(3) *Which account of persistence should dispositionalists endorse?*
Again, the question is not what best fits an ontology of dispositions, but what is the best ontology overall.

On (2) and (3): I admit that it can be useful to recognise that particular ontological theses are consistent with, or require, or exclude certain others. If you had strong views about substances, for instance, or about presentism, and you discovered that these views required or clashed with an ontology that included powers, this might lead you in one direction or another: you might rethink your prior commitments or reject powers, or find powers endearing. My concern is that treating such problems in isolation is in general a bad idea.

(4) *Are there irreducible powers only in the microphysical realm or also in the macrophysical realm?*
Properties are borne by substances, and substances are simple. Complex objects, quasi-substances, can be said to possess quasi-properties. You can perfectly well talk of powers possessed by molecules, billiard balls, clouds, and planets, but truthmakers for truths about such things are going to be organised, interrelated, interacting collections of propertied simples (if the particles are fundamental) or ways a field, or space, or space-time is (if the substances are not particles).

I realise that many will balk at these answers. This is philosophy after all. You have to start somewhere, however, then make adjustments. I feel certain that many of you are itching to tell me where I must adjust.

References

Armstrong, David M. 1978. *Universals and Scientific Realism*, 2 vols. Cambridge: Cambridge University Press.
———. 1989. *Universals: An Opinionated Introduction*. Boulder: Westview Press.
———. 1997. *A World of States of Affairs*. Cambridge: Cambridge University Press.
Campbell, Keith. 1990. *Abstract Particulars*. Oxford: Basil Blackwell.
Heil, John. 2003. *From an Ontological Point of View*. Oxford: Clarendon Press.
———. 2012. *The Universe as We Find it*. Oxford: Clarendon Press.
Huemer, Michael, and Ben Kovitz. 2003. Causation as Simultaneous and Continuous. *The Philosophical Quarterly* 53: 556–565.
Ingthorsson, Rögnvaldur D. 2002. Causal Production as Interaction. *Metaphysica* 3: 87–119.
Marmodoro, Anna. This volume. Powers, Activity and Interaction. In *Dispositionalism. Perspectives from Metaphysics and the Philosophy of Science (Synthese Library* 417*)*, ed. A. S. Meincke, 55–66. Cham: Springer.
Martin, Charles B. 2008. *The Mind in Nature*. Oxford: Clarendon Press.
Meincke, Anne Sophie. This volume. Powers, Persistence and Process. In *Dispositionalism. Perspectives from Metaphysics and the Philosophy of Science (Synthese Library* 417*)*, ed. A. S. Meincke, 89–113. Cham: Springer.
Mumford, Stephen, and Rani L. Anjum. 2011. *Getting Causes from Powers*. Oxford: Oxford University Press.

Ryle, Gilbert. 1949. *The Concept of Mind*. London: Hutchinson and Co.
Shoemaker, Sydney. 1980. Causality and Properties. In *Time and Cause*, ed. P. van Inwagen, 109–135. Dordrecht: Reidel. Reprinted in his *Identity, Cause, and Mind: Philosophical Essays*, 1984, Cambridge: Cambridge University Press, 206–233.
———. 1998. Causal and Metaphysical Necessity. *Pacific Philosophical Quarterly* 79: 59–77.
Williams, Neil E. This volume. What Are Manifestations? In *Dispositionalism. Perspectives from Metaphysics and the Philosophy of Science (Synthese Library* 417*)*, ed. A. S. Meincke, 67–87. Cham: Springer.

Chapter 3
What Does the Doing? On Powers, Things and Powerful Things

Ruth Groff

Abstract Assuming a neo-Aristotelian anti-passivist metaphysics, I pose the question "Is it powerful things that produce outcomes, or the powers of powerful things?" I argue that it is powerful things, not the properties thereof, that 'do the doing.' This claim turns out to be connected to several others. Thus, in the course of advancing the main thesis of the chapter, I also suggest: (i) that we must take care not to conflate causation with causes; (ii) that objects (or phenomena generally) and their (essential) properties do not exist separately from one another; and (iii) that a trope-theoretic reduction of ostensibly-material entities to collections of abstract particulars is harder to defend if one is not a Humean.

Keywords Powers · Causation · Properties · Causal dispositionalism · Passivism · Anti-passivism · Neo-Aristotelianism · Tropes · Jonathan Lowe · Roy Wood Sellars

3.1 Introduction

In arguing for the existence of real causal powers, those who believe in them have been concerned to show that the situation is not as the Humean, or the passivist (a term that is both more general and more precise), would have it, when it comes to causation.[1]

This article is dedicated to Jonathan Lowe.

[1]A thinker who identifies as a Humean but who believes in the existence of real and irreducible causal powers, conceived in terms of real and irreducible productive activity, will not be covered by my use of the label 'Humean.' Conversely, a thinker who defends the existence of something that s/he calls a 'power,' but for whom the term does not refer to the ability had by a thing to engage in real productive activity of a given kind, will not count as holding the anti-passivist position that the following argument presupposes. Alexander Bird, for example, who contends that there is no non-metaphorical sense in which the world is or contains "activity," counts as a passivist in this

R. Groff (✉)
Department of Political Science, Saint Louis University, St. Louis, MO, USA
e-mail: rgroff@slu.edu

A. S. Meincke (ed.), *Dispositionalism*, Synthese Library 417,
https://doi.org/10.1007/978-3-030-28722-1_3

Regularities; counterfactuals; laws of nature; being the case in all possible worlds – none of these phenomena bring about change, according to the anti-passivist. Powers do. But once we are no longer battling it out with passivists, a different question comes to the fore. Is it really properties, here powers, that do the doing, when a change is brought about? Or would it be better to say that the 'doing' is done by the *bearers* of powers, things ("thing" as a count noun only) that are powerful, but are not themselves powers?[2] One kind of answer is given by Stephen Mumford and Rani Lill Anjum in *Getting Causes from Powers*: "Weight has to be the weight of something. [. . .] So it is quite reasonable to say that the apples moved the scales" (2011, p. 1). Yet "[i]t is [. . .] *something about* the substance that does its causal work" (2011, p. 2). Specifically, "[it] is properties that do the causal work, and they do so because they are powerful" (2011, p. 1). The notion that powers *themselves* are powerful has gained traction in the course of recent debates about properties, i.e., in the argument had with categoricalists over whether or not powers exist at all, and the subsequent in-house disagreement amongst anti-passivists over pandispositionalism. In the wake of such disputes, propertied things seem to have fallen by the wayside.[3] Friends of powers, and even foes, have come to conceive of powers as efficacious entities in their own right, in contrast to categorical properties, which (it is sometimes said) are phenomena that can be, but not do.[4] Still, as Mumford and Anjum observe, it *is* reasonable to say that the apples moved the scale. Thus I am going to press the point. What actually does the

context – as does Spinoza, for instance, for whom a power is a relation of conceptual necessitation, not an irreducible capacity for activity, dynamically construed. (For a longer discussion, see Groff 2013). Contemporary thinkers who connect a belief in causal powers to a rejection of Humean and neo-Humean accounts of causation include, most notably (though not exclusively): Brian Ellis, Stephen Mumford and Rani Lill Anjum, Rom Harré and E. H. Madden, Roy Bhaskar, and Sterling Power Lamprecht. Nancy Cartwright, John Heil, Anjan Chakravartty, and G. E. M. Anscombe may also be included in this group.

[2]I do not mean to be weighing in on the debate over "pure powers". The question is whether or not the properties of a thing, rather than the propertied-thing itself, are what does the doing, not whether or not categorical properties exist, and if they do whether or not dispositional properties, or powers, in any way require or presuppose them. This said, the tendency to begin to view powers themselves as being powerful particulars is perhaps a greater risk for pandispositionalists than for others.

[3]Nancy Cartwright is an important exception.

[4]This locution is used in Ellis (2012), but Ellis is clear, even in the same paper, that properties are borne by substances, and that dispositional properties are properties related to what their *bearers* can do (categorical properties, he thinks, are properties that are *not* related to what their bearers can do; or if they are, then only indirectly as "dimensions" or "factors", occasions or contexts for the display of properties that *are* causal powers). But the literature is rife with those who believe that powers themselves are powerful. Sharon R. Ford, for example, writes: "Properties that are powerful might be best described as those that both *are*, and yet *do*, merely by virtue of doing" (Ford 2012, p. 187). Similarly, Stathis Psillos has criticised pandispositionalism precisely on the grounds that a belief in powerful powers invites a regress (see Psillos 2006). Anna Marmodoro has ably responded to this criticism; see footnote 8, below. An exception, again, is Bird, who does not think that the difference between dispositional and categorical properties (the existence of which he rejects) is that the former can do things; rather, he thinks that what he calls 'potencies' are properties that have an essential identity, while categorical properties are properties that (were they to exist) would have no essential identity.

doing? Is it (a) powerful things; or (b) the properties of powerful things? It will not do to try to have it both ways.

My view is that it is propertied things that are causal 'doers', not their properties, and I shall offer some reasons for thinking so in a moment. The thesis is less circumscribed than it might seem, however. Thoroughly prosecuted, it raises: (i) a distinction to be underscored between causes and causation; (ii) a unity to be affirmed between things and the ways that things are; and (iii) an implication to be appreciated concerning trope theory. The discussion to come is therefore organised into two parts. I begin by offering some grounds for thinking that the doing is done by powerful things, rather than by the powers of powerful things; then I address the points that I have just adumbrated. While the analysis has implications for the anti-passivist who may be attracted to the idea that material objects are actually bundles of tropes, it is not directed to those who begin by endorsing such a reduction. I assume a commitment to the existence of irreducibly material propertied things, and an associated conceptual distinction that may be made between propertied things and the properties of propertied things.

3.2 Powers Versus Powerful 'Things'

Powers, insofar as they are properties, are abstractions.[5] One might think that it goes without saying that abstractions, being abstract, cannot produce outcomes. But it seems plausible to think that some abstractions, anyway (belief systems; novels; the aesthetic content of a work of art, say), do affect outcomes – perhaps even *effect* them. It follows from this qualification that it will be necessary to distinguish the question of (a) whether it is powerful things that are the doers rather than the powers of powerful things from the question of (b) whether or not abstractions may be considered to be powerful things in their own right. (Again, "thing" is being used as a count noun only; the nature of the denoted entity does not figure into its being counted.) In the course of arguing that it is powerful things that are the doers, I am going to assume (though nothing hangs upon it) that some abstractions, at least, *may* well be powerful 'things.' The question at hand will obviously extend to such abstractions, if there are any: viz., is it they or their properties that do the doing?

As a matter of terminology, I use "powerful particular" interchangeably with "powerful thing," "propertied thing" and, on occasion, "causal bearer." The words "particular" and "bearer," are, in context, functioning along with "thing" as count nouns only. I shall assume for sake of discussion that events, too, as well as spatially disaggregated entities such as families and institutions, may be powerful particulars, along with some kinds of abstractions.

[5]Even as abstract *particulars* they are not themselves material entities. See the discussion of substances and their natures below. My thinking about this issue has been influenced by E. J. Lowe, though I do not claim that he would endorse my views.

Here are some grounds on the basis of which an anti-passivist might conclude that it is powerful things, and not powers themselves, that do the doing, as I shall continue to put it. We can begin with an indirect point. The best reason for holding the contrary position, the 'powers-as-doers' view, is that it does seem as though powerful particulars are only powerful in virtue of having the powers that they do indeed have. Manifestly, an allegedly powerful particular that had no powers would not actually be powerful. But this does not establish that it is the powers themselves that are the doers. If nothing else, the formulation cuts both ways: in saying that a thing is powerful, albeit in virtue of its properties, one may still be understood as having affirmed that *it*, i.e., the thing, is powerful – even if it is only powerful *qua* being the kind of thing that it is. And one might not (even) feel the need to add the "even." "All the same," the rejoinder may be, "if we were to take away its powers-to-do, there would be no doing." In responding in this way, the powers-as-doers proponent may presume herself to be simply subtracting variables, so as to home in on the source of the efficacy. John Stuart Mill called such an operation the "regulating principle" of the Method of Difference (Mill 2006, p. 391), and certainly those trained in the wake of Locke's introduction of the notion of a referent-less 'substrate' are accustomed to making use of the heuristic of a bare particular. However, a powerful particular stripped of its powers would not only be unable to *do* anything; it is not clear that it would be able to exist at all. The situation is maximally problematic if we assume pandispositionalism. If all properties are powers, a powerless particular would no longer *be* anything, since it is not possible to be a determinate thing that is no way at all. At least, it is not a possibility that is allowed by a neo-Aristotelian, who will insist that *all* things are *something*.[6] Not all anti-passivists are Aristotelians, of course, but those who are will balk at any argument concerning what does the doing that presumes the existence of property-less entities. Even the formal property of self-identity (a dialectical neo-Aristotelian will say), if it is to be a property *of* something, presupposes, for its application, the existence of a determinate 'this' that can be identical with itself.[7] Thus, although Mill's "were it not for" test may be diagnostically useful in some cases, the phenomenon in question does not appear to be one that admits of the relevant subtraction – not unless one is already committed to the dubious notion of a thing that is no way at all. If we assume only that *some* properties are powers, we are still left with a purportedly determinate entity that lacks all capacity to act or react. Again, the neo-Aristotelian anti-passivist will balk outright. But even if one were not committed to the view that to be is to be capable of one or another type(s) of activity, the idea that there are completely inert objects – or even objects none of whose powers are essential to them – is at odds with anti-passivism generally. And besides, the powerful-particulars-as-doers proponent

[6]It is not only the neo-Aristotelian who will balk. John Heil, for example, makes this point emphatically; see, e.g., Heil (2012).

[7]I do not mean to reject the concept of a *haecceity*. I do, however, mean to suggest that a particular that had no qualitative or substantive properties would also lack even a formal property of being itself. Of course, I have merely asserted this claim here, not argued for it.

can easily construct her own counterfactual appeal: were there no propertied things, there would just as certainly be no doing. An alternative strategy that the powers-as-doers proponent might pursue would be to say that the "were it not for" claim indicates an instance of counterfactual dependence, rather than a subtraction. The idea would then be that powers do the doing *inasmuch as* had there been no powers, there would have been no doing. But this approach will not fare well either, because the anti-passivist rejects the idea that counterfactual dependence is the stuff of causation. If it is true that it is powers rather than powerful particulars that do the doing, then it will certainly be true that there will be no doing in the absence of powers. But once we have rejected a conditional analysis of causation, this fact cannot be used to establish that it *is* powers that do the doing.

Second, while not all doing is what we might call "brute" doing, some is. And it is hard to imagine that properties, being abstractions, are the type of thing that can bang into other things, causing them to bruise or break. If it were so that it is properties that do the doing, and not propertied things, it would be difficult to account for doing that *is* brute, e.g., for breaking and bruising and banging. At least in *those* cases, it would seem, we will need something other than abstractions for the job of the productive causation the reality of which all parties to this discussion defend. Moreover, even if being an abstraction does not preclude a thing's being endowed with powers, it seems unlikely that powers would be amongst those abstractions that are. This because, as Anna Marmodoro has argued, powers don't *bear* powers; powers *are* powers (Marmodoro 2009).[8] Let me proceed very carefully here, lest there be confusion. The claim that I have just made is that an abstraction is not the kind of 'thing' that can be used to bang non-abstract nails into non-abstract wood. But I have also reiterated that, in my view, the mere fact of being an abstraction may not preclude an abstraction from being a bearer of powers – even if, *qua* propertied thing, an abstraction could not do what hammers (in conjunction with human beings) can do. This, as I have also just noted, then raises the question of whether or not powers as such are amongst those abstractions that one might think could be counted as powerful 'things.' And again: I am inclined to say no. There is a difference that makes a difference between properties and propertied-things-that-happen-to-be-abstract or conceptual. Properties are not themselves propertied things; properties just *are* properties, i.e., the properties of propertied things.[9] In the case of powers,

[8]Marmodoro argues against Stathis Psillos (see footnote 4) that pandispositionalism does not lead to a regress due to powers needing powers in order to be powerful. She draws a parallel between Psillos' claim and the issue of the relationship between entities and their essences in Aristotle's *Metaphysics*, suggesting that powers *are* powers in the same sense in which an entity is the instantiation of its essence. Her paper is therefore pertinent to this one in a second way, in that I argue below that the proponent of the powerful particulars-as-doers view will naturally reject any separation of things and their essential properties, here assumed to be powers.

[9]A trope theorist may deny that properties are borne by things that are not themselves clusters or bundles of properties. But (a) thinking that nothing exists but tropes does not necessarily commit the trope theorist to thinking that objects, conceived as nothing but tropes, do not have properties, conceived as tropes; indeed, the claim that properties are tropes rather than universals generally starts from at least a tacit presumption that properties (conceived as tropes) are properties *of* things;

they are the activity-related properties of powerful things. I will return to this issue in the second part of the chapter, but for now I want to note that even if one were to maintain that properties are no different from other abstract phenomena, such that properties, too, along with the content of novels, may be counted as being powerful particulars, thinking so would not settle the question of whether it is they or their own properties (now that one is allowing propertied properties) that do the doing. Note, however, that not only would such a view not settle the question (from which it follows that in denying it I am not begging the question), it has the weakness of affirming the very regress of propertied properties that Marmodoro shows (and which I will therefore not rehearse here) is *not*, contra Psillos, generated simply by affirming realism about causal powers.[10]

Third, the idea that it is powers that do the doing, and not the bearers of powers, invites special reservation when it comes to the theorizing of human agency. A decided advantage of an anti-passivist metaphysics with respect to the phenomenon of agency is that it naturally allows for agents, like other powerful particulars, to be taken to be causes. This feature of the metaphysics resolves not just well-known complaints about the improbability of agent-causation as a mysterious or *ad hoc* type of causation relative to event-causation, but also the intelligibility or traction problem faced by event-causal libertarians and compatibilists alike, viz., the question of how it can be that we are the authors of our acts, when these are thought to be either uncaused or caused by antecedent events, rather than by us.[11] The realist about powers who maintains that it is powers rather than powerful particulars that do the doing will forfeit the solution to the intelligibility problem that is afforded by a generalised commitment to anti-passivist substance-causation. She will be no better off than the passivist event-causalist when it comes to attaching causation to the free agent, for now it will be powers that turn out to be the 'agents,' rather than persons.[12]

The powers-as-doers proponent may counter that agents just *are* powers. One version of this response would go all the way down, bottoming out in the thesis that all things, and not just agents, are metaphysically equivalent to a plurality (or perhaps a totality) of properties, here powers to *phi*. Another version might have it that *bodies* are not equal to powers, but selves are. I shall return to the question of whether or not anti-passivism combines well with trope theory, but even at this stage in the argument we can appreciate that it is unlikely that either version of the claim that agents just *are* powers will solve the problem of the impotent, causally irrelevant agent. The problem arises in the first place because the powers-as-doers proponent is in the position of having to say that it is the agent's properties (here, powers) that do

---

and (b) the present discussion is not directed to someone who denies that there is a distinction between properties and propertied things. For a nice non-Aristotelian discussion of the relationship between the conceptually distinct (but concretely inseparable) categories of property and substance (or object) see John Heil (2012), though Heil believes that only simples have genuine properties (or, are genuinely propertied).

[10]See footnotes 4 and 8.

[11]Groff (2019). See also Groff (2013).

[12]Thanks to John Greco for this point. Public conversation, October 2012.

the doing, rather than the agent. In an effort to retain a causal role for the agent, she then says that there is no need to worry because agents are *composed* of powers. But this will not be enough to assign the agent a causal role, since the claim was that it is not she who does the doing. Relative to such a claim, the belief that the agent is made of powers is neither here nor there. "There *are* no agents who are the bearers of powers", one would have to say, in order to meet the objection; "there are just powers". Powers and powers and more powers. But once we have gone this far, it is not clear that we have succeeded in preserving a role for the agent after all, since now we no longer *have* an agent. Moreover, even if we were to grant our interlocutor an agent who could be said to exist but just *be* the very powers that are said to be "hers", agents comprised solely of abstractions would be as limited in their efficacy as would be any other abstract entity.

Fourth, in addition to making it hard to conceptualise agency, the powers-as-doers view does not actually sit well with a comprehensive anti-passivism. As noted earlier, recent work on powers has gravitated toward questions about the nature of properties, rather than focusing on the nature of propertied things. It is worth being reminded, therefore, that the fundamental thesis of a powers-based metaphysics is that things in the world are irreducibly such that they have a potential to be active. In principle, certainly, such a thesis is consistent with the view that it is only properties that are or can be active. The powers theorist is not obliged to think that every single spot of being is potent. But a picture of the world as consisting of inert propertied-things plus dynamic properties is not the one that coheres most easily or intuitively with a general commitment to anti-passivism. On the contrary, such a conception arguably preserves more of what Brian Ellis calls "the dead world of mechanism" than one might like (Ellis 2002, p. 60).[13] The powers-as-doers proponent may be able to avoid this outcome by opting (once again) for the view that deep down everything just is some number of properties (or clusters, bundles or sets of properties). Doing so, it would appear, would allow her to retain an ontology in which dynamism could be said to saturate all of being, assuming that she thought that all properties are powers. But if recourse to trope theory is the only way to salvage a comprehensive anti-passivist position, then we will need to be sure that trope theory itself can be plausibly combined with a powers-based account of causation. If there are reasons to doubt the cogency of such a pairing, then the bifurcated passivism/anti-passivism that is otherwise entailed by the powers-as-doers view will continue to count against the idea that it is powers rather than powerful things that do the doing. I shall pursue the issue below. For the moment, it is worth registering that it is an added cost of the powers-as-doers view that it looks to require for its defence that it be backed up by trope theory, even if such a combination turns out to be viable.

Finally, the idea that powers could be doers all by themselves presupposes a separation between properties and propertied-things that is itself a product of post-Aristotelian hostility to the notion of substantial form. To be sure, the fact of the

[13]Ellis credits E. A. Burtt's *The Metaphysical Foundations of Modern Physical Science*, published in 1932, for the metaphor. Brian Ellis, personal correspondence, March 2012.

genealogy does not decide the issue. Perhaps there really are featureless *substrata*, upon which powers alight, affording us the opportunity to proclaim or deny the potency of the properties themselves. But there are reasons to be sceptical of this supposition, especially if one is already sympathetic to Aristotle. Admittedly, the question as I have posed it here is whether it is powers or powerful propertied things that do the doing, not whether it is powers or featureless *substrata*. All the same, the very idea that it might be properties that are the productive causal doers (rather than propertied things) presupposes a metaphysics in which properties make their way in the world un-tethered by propertied things. Thus we should be under no illusion that the powers-as-doers position commits one to nothing beyond that which has been affirmed simply by believing in the existence of powerful particulars.

3.3 Regarding Powerful Particulars

Let us assume, now, that it is indeed propertied things, rather than their properties, that do the doing in causal processes, and consider several points that follow upon such a claim – if not as a matter of logical entailment, then at least as a matter of ampliative good sense. As we will see, at least one of the points to come was touched upon in the preceding analysis. The fact that it figures in the argument both coming and going, as it were, suggests that a neo-Aristotelian account of powers may be fundamentally connected to a neo-Aristotelian account of objects. It would not be a surprising discovery, if something like this were to turn out to be the case. Or at least it ought not to be. There is no special reason to think that an ontology must be built up only one conceptual building block at a time, and/or that any given block may be joined together with any other.

3.3.1 Causation and Causes

How should those who define causation in terms of the display of powers look upon the idea that it is powerful things, rather than the powers of powerful things, that do the doing? Mumford and Anjum's recent treatment of what they call "causal dispositionalism" is entitled *Getting Causes from Powers*, after all. Can we defend causal dispositionalism without contending that it is properties rather than propertied things that are causes? Of course we can. However, we must be alert to a potential conflation. Mumford and Anjum (2011) argue that causation occurs when powers that are "manifestation partners" come together: when two or more such powers meet up, their potentials for doing are expressed. For this reason, Mumford and Anjum suggest, causation is not best thought of as a relationship between a cause and an effect. Rather (they say), we should think of causation as a display of powers, a pragmatically delineated instance of activity. What happens at any given moment is a function of all of the expressed powers at that time, a scenario that Mumford and Anjum depict in the form of a vector model. An 'effect' is simply an indication that a threshold, as they put it, has been reached, i.e., that vectors of interest have reached a designated metaphorical line. Brian Ellis, similarly, talks about causation in terms of process kinds, which he defines as species of displays of dispositional properties (see, e.g., Ellis 2001). What both of these approaches correctly pick out is the dynamism, or activity, that is essential to causation. In fact, we might want to say that causation just *is* dynamism. But we must be careful not to elide the concepts of causation and cause. Causation is doing. Causes are that which can do. If one thinks that it is propertied things that do the doing, then one will be obliged to be diligent in maintaining the distinction. Causation may be the display of powers, but causes are propertied things. This said, there is nothing troublesome about the requirement. Above all, the stipulation that (a) it is causation, not causes, that is to be got from powers (if anything is); and (b) that while it is true that causes are causes in virtue of being powerful, they are neither powerful nor causes in virtue of *being* their powers – this stipulation in no way undermines a belief in maximally robust powers, or in an anti-passivist account of causation.

3.3.2 Powerful Particulars and Their Properties

If one is an anti-passivist who thinks that it is powerful particulars that do the doing, then one will be likely to object to any division between a thing and its properties that would permit either term to be thought to exist apart from the other. To be sure, one might have come independently to decry such a division, and having done so would be reason enough to reject the powers-as-doers view, since one who had already denied that there are property-less substrates would take powers construed as doers to be hypostatisations. But we can get to the view that things and their properties are unified from the other direction, too. If it is powerful particulars that do the doing, and they do so in virtue of being what they are, then it will natural to think of things and their properties as being integrated powerful particulars. Powers, one might want to say, are just ways that a powerful particular *is* – and, to reiterate, there is no compelling reason to imagine that there is such thing as an entity that is no way at all. Roy Wood Sellars expressed this thought beautifully, in an important if now largely forgotten article that appeared in *Mind* in the fall of 1929, entitled "Critical Realism and Substance". Properties, he wrote there (Sellars 2008, p. 15),

> are elements of the nature of the object. But even this way of putting it is not penetrative enough. We must not think of the nature of the object as in any way distinct from the object. The object is a *determinate* object. [. . .] To say that a thing has a definite structure does not mean that there is some substratum which owns an entity called structure but simply that the thing is intrinsically structured. [. . .] We speak of a thing (subject of the judgement) which has properties (predicates of the judgement). But we should penetrate beyond this logical form . . . Logical structure must not be identified with ontological structure.

E. J. Lowe (2012, p. 247), though friendlier to Locke than was Sellars, makes a similar claim about objects and what he calls their "modes":

> The proper thing for a four-category ontologist to say about modes, I believe, is that they are 'abstractions' from objects [. . .] I suggest, then, that [. . .] we need to regard modes as being 'aspects' of objects, to which we can attend selectively in thought or perception [. . .] These aspects explain the differential behavior of different objects: for instance, why some objects roll down an inclined plane while others do not – the former being *spherical* or *cylindrical*, the latter not. But it is, after all, *the whole object* that rolls, not its sphericity or cylindricality. Nor does it make much sense to suppose that its sphericity or cylindricality 'drags along' the object's other modes with it, and thereby makes the object as a whole move. To think in such terms is illicitly to hypostatize modes, treating them as simple substances *within* an object, rather than just a particular 'ways' an object is.

And John Heil, too (2012), holds this same view with respect to the relationship between a propertied thing and its properties. The point was originally Aristotle's, though. As Anna Marmodoro reminds us, Aristotle argues in the *Metaphysics* that "the instantiated essence of an entity is one and the same as the entity" (Marmodoro 2009, p. 348). To quote Aristotle himself: "Therefore it is clear that the causal responsibility attributed to the forms, in the sense that some people are in the habit of speaking of the forms, as if they are certain things apart from the particulars, is of no use, at least in relation to coming into being and independent things; nor would they be, for the sake of these things at least, independent things in their own right" (Aristotle 2002, *Metaphysics* VII. 8, 1033b 20–30).

It is not obvious that one *must* affirm the unity of powerful particulars and their powers, if one is a realist about powers who believes that it is propertied things that are the doers. But it is the stance that it will make easy sense to adopt, if one does hold that view. I say this because the alternative view (viz., that powerful things are really bare particulars to which powers have been added) implicitly commits one to a belief in denuded phenomena that manifestly are *not* doers. No doubt a presumption of the reality of bare particulars, held amongst anti-passivists not strongly influenced or informed by Aristotle, does tacitly invite support for the powers-as-doers view.

3.3.3 Powerful Particulars and Trope Theory

Earlier, I noted that one might defend the view that it is powers that do the doing by saying that powerful things just *are* powers, or compresent sets or clusters thereof. Let me specify that I have in mind trope theory of the D. C. Williams variety (Williams 1953a, b), not a view according to which the constitution of objects also involves a so-called kernel. Tropes, says Williams, "are the primary constituents of this or any possible world, the very alphabet of being. They are not only actual but are the only actualities, in just this sense, that whereas entities of all other categories are literally composed of them, they are not in general composed of any other sort of entity" (Williams 1953a, p. 7). And again "[a]ny possible world, and hence, of course, this one, is completely constituted by its tropes and their connections of location and similarity, and any others there may be" (Williams 1953a, p. 8). If one holds the view that it is powerful things rather than their powers that do the doing, and one's reasons for thinking so include a belief that abstractions cannot be efficient causes in cases of what I called above "brute" doing, then one is probably unlikely to think that powerful things just *are* powers. Nonetheless, the possibility of an appeal to trope theory by a pandispositionalist powers-as-doers proponent forces us to reflect more generally on the relationship between trope theory and anti-passivism. To what extent may these foundational metaphysical theses be combined?

This much, I think, will be fine: the abstract particulars that (according to the classical trope theorist) constitute all other entities need not be categorical properties. They can be powers. It would therefore appear to be an option for the trope theorist to subscribe, at least provisionally, to a powers-based account of causation. It is true that Ellis, for example, believes in *kinds* of powers, which are universals, but this alone will not preclude the trope theorist from endorsing even an Ellis-style anti-passivism. She will simply need to specify that universals are to be defined in terms of similarity. Nor will there be an immediate problem for the trope theorist who prefers Mumford and Anjum's approach. There is nothing to stop her from agreeing that when the powers of manifestation partners are expressed, causation occurs – and there is no reason why she may not also be a pandispositionalist, if she likes.

But there will be trouble when it comes to question of what does the doing – especially so if our trope theorist is a pandispositionalist. Again, insofar as objects, for the trope theorist, *just are* clusters or sets of abstract particulars, as Williams puts it, it will be in the nature of the case, for her, that it is properties that do the doing when causation occurs – since properties (now powers exclusively) are all that there is, in the end.[14] She will thereby hold a position that one might be tempted to call 'powers-causation', rather than substance-causation. I have already suggested that there is some tension between *any* version of the powers-as-doers view and anti-passivism, in that the powers-as-doers view leaves us with a dichotomy between active properties and the inert, causally superfluous propertied things that turn out not to be powerful. There was the hope, in the earlier iteration of the problem, that trope theory could resolve the tension (at least for entities in relation to which we need only establish that they are active, and not that they exercise agential control over their capacities). What we need to determine, therefore, is whether or not there is further tension between an anti-passivist metaphysics and trope theory *specifically*, tension that would preclude trope theory being an answer to the passivism implied, ironically enough, by the powers-as-doers position. And the answer, I think, is yes. There is.

[14]The trope theorist may object that she does not mean to eliminate substances. Substances exist; it's just that they are clusters of properties. But either the trope theorist believes (non-trope) substance ontologies to be mistaken or she doesn't. Either properties are the alphabet of being or they aren't. Substance theorists and trope theorists disagree on this point, and for better or for worse one can't have it both ways.

Here's why. The trope theorist has, as a general task, the challenge of showing the non-trope theorist who is a materialist how it is that clusters or bundles of abstract particulars add up to things that behave in the ways in which ostensibly non-abstract things behave. Note that this is a challenge faced by all idealists, i.e., by all who assume that the "alphabet of being" is comprised entirely of abstract or conceptual phenomena, just as linguistic alphabets are. Note too that the materialist has more or less the opposite problem: viz., how to get consciousness from that which is material (i.e., consciousness understood as something that is not identical to that which is material, either at the level of property or at the level of substance). Materialists often draw upon the concept of emergence in order to account for the existence of material substances capable of cognition. To the best of my knowledge, trope theorists have not handled the problem of getting material objects from abstractions by an appeal to emergence. Rather, the classical trope theorist argues that ostensibly material, non-abstract entities – middle-sized dry goods, as the phrase goes – just *are* collections of *abstracta*, not that they are emergent phenomena vis-à-vis bases consisting solely of *abstracta*. They are composed of abstract *particulars*, to be sure, but they are composed of abstractions all the same. And the point is this: such an idea will be a considerably tougher sell to the anti-passivist than to the Humean. Why? *Because in a Humean ontology, middle-sized dry goods do not have to actually do anything.* They just have to always come first in regular sequences. The trope theorist might respond to this latest form of the objection that it is hard to imagine a cluster of abstractions acting as an efficient cause, as I put it earlier, precisely by reverting to the powers-as-doers view: we do not need our trope-constituted powerful particulars to be able to bruise or break things, because it is not powerful particulars that do the doing. Doing is done by the powers of powerful particulars. But if there is reason to doubt that a cluster of properties can be a brute doer *qua* middle-sized dry good, there will be the same reason to doubt that clusters of properties can bruise, break or bang into things *qua* the powers thereof – as previously argued.

Seth Shabo observes that the trope theorist may say that there is no problem because the nails, for example, to which a hammer composed entirely of abstractions would have to impart force are also made of abstractions.[15] But this just brings us back to where we began: the trope-theorist really does have to be able to explain how it is that entities made of abstractions can do things that we otherwise believe cannot be done by abstractions. How is it that a hammer composed entirely of *abstracta* can do something that all are likely to agree cannot be done by the concept of a hammer – or, for that matter (at it were), by a painting of a hammer? My claim was that the challenge faced by trope theorists (viz., of accounting for the behaviour of ostensibly-material objects) becomes more difficult if the trope theorist is not a passivist. Pointing out that from an idealist perspective the nail, too, is made of abstractions, is just as likely to weaken her case as it is to strengthen it. The upshot, I think, is that it is not at all obvious that trope theory can sustain an anti-passivist

[15] Personal conversation, 19th October 2015.

account of causation. By requiring causes that can bruise, break and bang, a powers-based approach puts materialist pressure on the trope-theoretical analysis of presumptively-material objects that Humeanism does not.

Finally, while a powers-as-doers proponent might have looked to trope theory in an effort to defend the idea that it is powers that do the doing, the reverse is also possible. She may have endorsed the powers-as-doers thesis so as to accommodate a commitment to trope theory. But if I am right that trope theory and a powers-based account of causation are not ultimately compatible, then it will be a bad idea to plump for properties rather than propertied things as doers simply to try to make causal dispositionalism tenable to trope theorists. Nor will it make much sense to claim agnosticism about what does the doing, or to say that it is sometimes powerful things, other times powers on their own. One can certainly say that one has not made up one's mind, but I do not see that there is a way to be neutral on the question, or to have it both ways. The problem with saying that sometimes it is one way, other times the other way, is that if it really could be either way, then it is not clear why it would ever be the powerful-things-way. If powers alone could get the job done, the exception would be unmotivated. A better bet for (in some sense) having it both ways would be to argue that powers should be construed as limit cases of unified powerful particulars: properties would be the ultimate 'substance'-causes.[16] But the equation of powers with powerful particulars that is implicit in the move would then render unmotivated the alternate caveat, i.e., the stipulation that sometimes the doing is *not* done by powerful particulars (including, now, powers) but by the powers of powerful particulars.

Happily, there is no need to be alarmed. Nothing of consequence to the anti-passivist is threatened by the conclusion, which I hope to have shown to be compelling, that while causation is the expression of powers, it is propertied things that do the doing, not their properties.

References

Aristotle. 2002. *Metaphysics*. Trans. Joe Sachs. Santa Fe: Green Lion Press.

Bhaskar, Roy. 1978. *A Realist Theory of Science*. Sussex/Atlantic Highlands: The Harvester Press.

Ellis, Brian. 2001. *Scientific Essentialism*. Cambridge: Cambridge University Press.

———. 2002. *The Philosophy of Nature: A Guide to the New Essentialism*. Montreal/Kingston: McGill-Queen's University Press.

———. 2012. The Categorical Dimensions of the Causal Powers. In *Properties, Powers and Structures: Issues in the Metaphysics of Realism*, ed. A. Bird, B. Ellis, and H. Sankey, 11–26. New York/London: Routledge.

Ford, Sharon R. 2012. The Categorical-Dispositional Distinction. In *Properties, Powers and Structures: Issues in the Metaphysics of Realism*, ed. A. Bird, B. Ellis, and H. Sankey, 181–200. New York/London: Routledge.

[16]Rom Harré and E. H. Madden (1975) and Roy Bhaskar (1978) defend the view that ultimate entities are nothing but 'their' powers.

Groff, Ruth. 2013. Whose Powers? Which Agency? In *Powers and Capacities in Philosophy: The New Aristotelianism*, ed. R. Groff and J. Greco, 207–227. New York: Routledge.
———. 2019. Sublating the Free Will Problematic: Powers, Agency and Causal Determination. *Synthese* 196 (1): 179–200. (Special Issue: *Real Possibilities, Indeterminism and Free Will*). https://doi.org/10.1007/s11229-016-1124-y.
Harré, Rom, and Edward H. Madden. 1975. *Causal Powers: A Theory of Natural Necessity*. Oxford: Basil Blackwell.
Heil, John. 2012. *The Universe As We Find It*. Oxford: Clarendon Press.
Lowe, E. Jonathan. 2012. A Neo-Aristotelian Substance Ontology: Neither Relational Nor Constituent. In *Contemporary Aristotelian Metaphysics*, ed. T. E. Tahko, 229–248. Cambridge: Cambridge University Press.
Marmodoro, Anna. 2009. Do Powers Need Powers to Make Them Powerful? From Pandispositionalism to Aristotle. *History of Philosophy Quarterly* 26 (4): 337–352.
Mill, John Stuart. 2006. *A System of Logic, Racionative and Inductive: Being a Connected View of the Principles of Evidence and the Methods of Scientific Investigation, The Collected Works of John Stuart Mill, Vol. VII*, ed. J. M. Robson. Indianapolis: The Liberty Fund.
Mumford, Stephen, and Rani L. Anjum. 2011. *Getting Causes from Powers*. Oxford: Oxford University Press.
Psillos, Stathis. 2006. What Do Powers Do When They Are Not Manifested? *Philosophy and Phenomenological Research* 72 (1): 137–156.
Sellars, Roy Wood. 2008. *Critical Realism and Substance*. Reprinted in *Revitalizing Causality: Realism about Causality in Philosophy and Social Science*, ed. R. Groff, 13–25. London/New York: Routledge.
Williams, Donald C. 1953a. On the Elements of Being: I. *The Review of Metaphysics* 7 (1): 3–18.
———. 1953b. On the Elements of Being: II. *The Review of Metaphysics* 7 (2): 171–192.

Chapter 4
Active Bearers: The Ontology of Mental Dispositions

Josef Quitterer

Abstract In this chapter, I discuss the ontological implications of the assumption of dispositional mental states. To that end, I analyse the relationship between mental dispositions and their manifestations, looking at views of mental dispositions advanced by Gilbert Ryle, Alvin Goldman, Stephen Mumford and Rani Lill Anjum. I show that while on all these views it would be natural to assume enduring agents as ontological bearers of mental dispositions, there is no such explicit ontological commitment and neither to agent causation. The reason for this ontological abstinence is that causation is taken to be brought about by mental events or mental powers. I argue for the assumption of enduring bearers of mental dispositions and defend the claim that this commits us to agent causation and to acknowledging a fundamental difference between the manifestation of mental dispositions and that of physical dispositions.

Keywords Mental dispositions · Endurers · Mental causation · Agent causation

4.1 Mental Dispositions and Their Manifestations

Mental dispositions play a marginal role in the literature on dispositions; the focus is on physical dispositions. Do mental dispositions require a special treatment? What are the ontological commitments incurred by endorsing mental dispositions? The domain of mental dispositions comprises, *inter alia*, the following phenomena: epistemic virtues (Henderson and Horgan 2009), mental capacities such as the capacity to have a first person-perspective (Baker 2013), 'standing' reasons and beliefs (Goldman 1970; O'Connor 2009) and the power of willing (Mumford and

J. Quitterer (✉)
Department of Christian Philosophy, University of Innsbruck, Innsbruck, Austria
e-mail: Josef.Quitterer@uibk.ac.at

A. S. Meincke (ed.), *Dispositionalism*, Synthese Library 417,
https://doi.org/10.1007/978-3-030-28722-1_4

Anjum 2011). Mental dispositions, like all other dispositions or powers,[1] are linked intrinsically with their manifestations, that is to say, with specific activities or realisations which they are disposed to perform or manifest. The nature and identity of dispositions is determined through their manifestations: "Manifestations are isomorphic with powers because each power gets its identity from its manifestation" (Molnar 2003, p. 195). According to Lowe, it cannot be a merely *accidental* feature of a given power that it is a power to do such-and-such: rather, it must be a part of its *essence* (Lowe 2010, p. 10). Thus, before making any claims about the ontological commitments of the ascriptions of mental dispositions, we must determine their nature by clarifying the relation between mental dispositions and their manifestations.

Looking at the scholarly debate, there are two different ways in which mental dispositions can be regarded to manifest themselves. *First,* mental dispositions can be understood as mental states of which we are not actually aware, and which are manifested through mental representation. We assume the existence of intentions, beliefs, wants and other mental states, even when we are not actually aware of having them. For example, my intention to go skiing as soon as there is enough snow in the mountains can exist even if I do not think about it. The manifestation of these mental dispositions occurs when we have conscious access toward them. My dispositional intention to go skiing is manifested when I become aware of having the intention to go skiing – for example when my colleague askes me whether I have the intention to go skiing this afternoon.

Second, mental dispositions can be interpreted as intentional mental powers whose manifestations consist in actions or just in the exercising of the mental power. For example, my intention to go skiing as soon as there is enough snow can also manifest itself through the realisation of my intention – my skiing on the snowy mountains. In this case, the dispositional character of the intentional state consists in a mental power which is manifested in the corresponding behaviour. Or my willing power to lift the arm can be manifested through the exercising of the willing power – even if some circumstances prevent the execution of the act of lifting my arm.

In the following, I will provide a more detailed analysis of the different views about how mental dispositions can become manifest. I will then elaborate on the ontological implications of the respective relations between the dispositions and their manifestations, discussing the following questions: What are the best candidates for being the bearers of mental dispositions? Can there be event- or process-like bearers? Or do we need enduring entities? I shall argue for the latter and investigate the question of whether enduring entities play an active role in the manifestation of the dispositions as agents. If so, the causation involved in an agent's performing an action is not reducible to causation among events involving the agent. Rather, the agent has to provide an irreducible causal contribution to the manifestation of the disposition.

[1]Molnar seems to use the terms 'disposition' and 'power' synonymously (see Molnar 2003, p. 60). In my contribution, I shall apply the notion of a 'mental power' to mental dispositions insofar as their manifestations are taken to consist in the mental dispositions' being exercised or in actions.

4.2 Powerless Dispositions – Powerful Agents?

In the analysis of the first way of conceptualizing the manifestation of mental dispositions, I shall look at Alvin Goldman's distinction between "standing" and "occurrent" mental states (Goldman 1970, p. 86). Goldman's view is paradigmatic of the treatment of mental dispositions in classical *philosophy* of mind. The main focus lies on mental phenomena which have an event-like occurrent character, such as volitions, intentions, desires and beliefs. The assumption of mental dispositions is motivated mainly by the common-sense intuition that we do have these mental states even at times when we are not actually aware of them:

> An occurrent want is a mental event or mental process; it is a 'going on' or 'happening' in consciousness. A standing want [. . .] is a disposition or propensity to have an occurrent want, a disposition that lasts with the agent for a reasonable length of time. Though it is [. . .] possible for a person to have such a disposition without ever having any manifestations thereof, we would not ascribe such a disposition to anyone unless the relevant manifestations appeared from time to time. (Goldman 1970, p. 86).

What is the relationship between the mental disposition and its manifestation on this view? This question can be answered by a more detailed analysis of the circumstances in which, according to Goldman, a standing mental state becomes an occurrent one. This analysis shows that the relationship between the mental disposition and its manifestation is not causal: mental dispositions have no causal power. They themselves cannot cause acts. They "can affect action only by becoming activated, that is, by being manifested in occurrent wants and beliefs" (Goldman 1970, p. 86). This means that the disposition itself does not cause or bring about its manifestation; rather, the manifestation consists in the mental state *being activated* through an act of consciousness: my standing want to eat ice cream this afternoon becomes an occurrent want only if I am aware of my want to eat ice-cream (Goldman 1970, p. 88).

First-person access to my mental state is constitutive of occurrent mental states (Goldman 1970, p. 97, p. 116). Although Goldman denies that self-reflection or explicit introspective awareness is necessary for occurrent mental states, he maintains that some form of knowledge of one's mental state is constitutive of having an occurrent mental state: "having an occurrent want carries with it a sort of knowledge of one's want" (Goldman 1970, p. 97). Hence, a higher-order mental state (knowledge of one's want) is required for the mental state to become an occurrent one. In this way, Alvin Goldman's conception of the relationship between mental dispositions and their manifestation can be regarded as an early version of a Higher-order Theory (HOT) of consciousness. In the same vein as HOT approaches, the relationship between the mental disposition and its manifestation is regarded as representational. A standing mental state is transformed into an occurrent mental event by becoming the object of a higher-order representation.

What are the ontological implications of this conception of mental dispositions? It seems obvious that we need a subject, and that this subject must be numerically self-identical at least during the dynamic change from having a mental disposition to manifesting that mental disposition in a mental event. The subject S', who is the bearer of a mental disposition at t_1, must be numerically identical with the subject S", in whom the mental disposition is activated at t_2, and with the subject S"', who is the bearer of the mental event manifesting the mental disposition at t_3. Otherwise, each moment in the manifestation-process of the disposition would belong to a different bearer. We would have a subject S' with an unmanifested mental disposition, a subject S" with the property of activating the mental disposition of S', and a subject S"' who is the bearer of the final mental event. As a result, the relation between the mental disposition and the act of consciousness through which the disposition becomes manifest would not any longer be a sort of knowledge of one's own mental state. It would become an epistemic relation between an act of consciousness and another subject's mental state. Goldman himself seems to acknowledge the need for a numerically self-identical agent when describing a standing want as "a disposition that lasts with the agent for a reasonable length of time" (Goldman 1970, p. 86).

We might, however, wonder whether the enduring agent plays an active role in manifesting the disposition. In classical philosophy of mind, the emphasis lies on mental events as causes of actions – and not on enduring substances as agents who bring about their effects. Hence, though mental dispositions are attributed to the same agent, there is no agent causation in the proper sense of this term. According to Goldman, agents are causes insofar as they are involved in events, or insofar as there are mental events within the agent, who cause the agent to act in a certain way:

> [...] I contend that in the case of agents, as in the case of other objects, there is no incompatibility between saying that a certain agent was the (object-) cause of a certain event and saying that an event involving the agent was the (event-) cause of the same effect. [...]
>
> We can say, without contradiction, both that John's flipping the switch caused the light to go on and that John caused the light to go on. Similarly, I think, when an act of John's is caused by John's wants and beliefs, it would not be incorrect to say that John caused this act. (Goldman 1970, p. 82)

In Goldman's view, agent causation is not only compatible with "want-and-belief causation", but also "explicable in terms of" it (Goldman 1970, p. 83). Mental events in the agent are the causes of her behaviour. Why should the agent-subject play any causal role in the manifestation of the disposition? As we saw above, the mental disposition itself is not a mental power which would manifest itself provided nothing interferes. The mental disposition rather must be activated by an act of consciousness.

According to Goldman, assuming that mental dispositions exist only makes sense in connection with mental events. It is another (higher-order) mental event that leads to the manifestation of the mental disposition by causally affecting that disposition. The mental event of my thinking about my want to eat ice cream this afternoon is taken to causally affect my dispositional wish for ice-cream and trigger the manifest mental event, namely, my wanting to eat ice cream. Such higher-order thoughts, however, presuppose a capacity to think about a given mental state. The capacity to think about wants, desires, intentions or beliefs exists also when I am not actually engaged in any self-referential activities. This capacity is a genuine dispositional property that is manifested in higher-order mental events.

What are the ontological implications of this dispositional notion of higher-order reflection on first-order mental states? First, the capacity to think about mental states is a property of entities with specific cognitive capacities, i.e., subjects. As properties, these dispositions are ontologically dependent on those entities in the sense that the latter are their upholders or bearers. Second, these mental capacities are intrinsically connected to their manifestation. The manifestation of the agent's capacity to think about her own mental states is not a property or state of affairs extrinsic to the disposition itself. When the disposition is manifested via thinking, the manifestation is rather intrinsic to the reflexive capacity itself. If an agent has the disposition to think about her mental states (e.g., about the wish to eat ice cream) at $t_{1,}$ and the disposition is manifested (the agent actually thinks about her ice-cream-wish) at t_2, then the bearer must endure at least from t_1 to t_2. It thus seems clear that, also in the case of higher-order mental states, we need ontological 'upholders' which exist ('endure') as numerically the same entities at least as long as the disposition lasts.

At this point, the following question arises: who or what is causally responsible for the manifestation of the capacity of self-awareness? It is possible to assume additional – third-order – mental events which are causally relevant to the manifestation of second-order reflective capacities. For example, the manifestation of the capacity to think about my want to eat ice-cream could be triggered through a third-order representation of the second-order reflection of my desire for ice cream. However, this assumption requires, once more, a corresponding capacity to perform such higher (third)-order reflection; and this capacity once more requires an enduring entity as its ontological bearer – and so on. The easiest way to avoid an infinite regress is to attribute, to the enduring subject, the causal power to activate the mental capacity of reflecting on one's own dispositional mental states. It is because *I* reflect on my dispositional wish to eat ice cream that this wish becomes a manifest mental event.

In summary, the classical theory of powerless mental dispositions seems ontologically committed to a thinking subject who is aware of the mental dispositions and who is able to represent them from a first-person perspective. It is natural to assume that this subject must be numerically identical through time, at least during the dynamic change from *having* a mental disposition to *manifesting* that disposition in a mental event. Still, genuine agent causation is not endorsed.

4.3 Powerful Dispositions – Powerless Agents?

The second conception of mental dispositions construes these as standing in an active relationship to their manifestations. Mental dispositions produce their manifestations, which consist typically in an action or overt behaviour. Mental dispositions in this sense are mental powers. In what follows, I shall analyse this assumed active relationship between the disposition and its manifestation: what are the ontological implications of this understanding of mental dispositions? Does the existence of powerful mental dispositions – of mental powers – imply that the agents who have them are powerless? Does this understanding of mental dispositions even need agents as ontological bearers at all?

Let us first look at Gilbert Ryle's dispositional view of mental concepts. In his book *The Concept of Mind*, Gilbert Ryle provides a detailed analysis of the logical structure of our everyday mental ascriptions. According to Ryle, mental terms should not be understood as 'episodic', i.e., in terms of single manifestations of mental activities, but rather as 'dispositional', i.e., as expressing certain capacities or powers to bring about specific types of behaviour:

> To say that a person knows something, or aspires to be something, is not to say that he is at a particular moment in process of doing or undergoing anything, but that he is able to do certain things, when the need arises, or that he is prone to do and feel certain things in situations of certain sorts. (Ryle 1949, p. 116)

For Ryle, mental dispositions are not manifested in mental events, but in actions or behaviours: mental dispositions "signify abilities, tendencies or proneness to do things" (Ryle 1949, p. 118). Ryle characterises dispositional mental terms as being "like law statements" or "inference-tickets, which license us to predict, retrodict, explain and modify [. . .] actions, reactions and states". Although these statements seem to suggest there to be a causal mechanism linking dispositions and their manifestations, Ryle explicitly rejects this. He argues against the idea that dispositional mental states (such as wants or intentions) exist prior to the acts that are wanted or willed – they are not like causes which precede their effect in a "para-mechanical pattern" (Ryle 1949, p. 68). The relationship between a disposition and its manifestation is not one of temporal succession but rather a relationship between two simultaneously coexisting aspects of the action. When I say, for example, that an action was performed by an enormous effort of will, I am not referring to a special cause preceding the action; instead I want to express a "particular exercise of tenacity of purpose, occurring when the obstacles are notably great" (Ryle 1949, p. 73).

Ryle sharply distinguishes between natural causes and the mental "abilities to do things" (Ryle 1949, p. 118). He describes the former as mechanical, following a temporal order of cause and effect, and the latter as propensities or capacities that can be effective simultaneously with the realised manifestation. In contrast, Stephen Mumford and Rani Lill Anjum propose a model of causation in which the difference between mental and physical powers is only a matter of degree. On this basis, they argue for a simultaneity of cause and effect in the manifestation of both physical and mental dispositions.

According to Mumford and Anjum, the classical Humean model, which reduces causal relations to a regular sequence of temporarily successive events, fails to capture our common-sense intuitions about causal processes. The main aspect of the Humean view denied by Mumford and Anjum is the temporal separation between causing and resulting event. On their view, it is counterintuitive to suppose that the causing event stops at the moment when the resulting event begins. In response, Mumford and Anjum take powers to be the driving force of all causal processes. On their account, a caused effect is (normally) the result of a combination of powers which manifest themselves with varying degrees of intensity. The effect occurs when the powers, taken together, reach the specific threshold required for the specific effect in question:

> The threshold account of causal production states that an effect is produced when some local aggregation of operative powers reaches the requisite threshold for that effect. In other words, an effect is caused when powers have accumulated to reach the point at which that effect is triggered. (Mumford and Anjum 2010, p. 145)

Here no important difference is assumed between mental dispositions, such as the willing of something, on the one hand and natural physical dispositions, such as the solubility of sugar, on the other. Causation through mental dispositions is taken to be a special case of the more general causation by powers. Thus, Mumford and Anjum analyse causation through both mental and physical dispositions in terms of a combination of different powers, construing the effect as the result of an interplay between enhancers and interferers. Mental powers as desires "dispose an agent to act though they may not lead directly to action if they are not accompanied by appropriate beliefs and other suitable circumstances and indeed physical powers are also needed in order for an agent to act" (Mumford and Anjum 2015, p. 137). Sugar's physical disposition of solubility manifests itself in the dissolving of the sugar, provided certain other powers (such as the dissolving power of the water) are present and not being prevented from manifesting by any interfering powers.

Mumford and Anjum invoke the power of *willing to raise one's arm* as a paradigmatic case of a mental disposition or power (Mumford and Anjum 2011, p. 206). Although the willing-power is a necessary condition for the causal production of the action, it is not a sufficient one. This power can be effective only in combination with a set of other powers.

The action of raising one's arm can be regarded as a causal effect of a combination of specific powers, which together produce the action. For Mumford and Anjum this is clearly a causal relationship, but it is not causal in the sense favoured in classical philosophy of mind, according to which there is a relationship between a mental cause-event and its subsequent action-event. Rather, Mumford and Anjum's model of causation through powers is meant to avoid the problems of classical event causality. The main aspects of the classical view denied by Mumford and Anjum are the temporal separation of, and the assumption of a necessary connection between, cause and effect (Mumford and Anjum 2011, pp. 11, 48, 205). On their view, it does not make sense to explain the arm's lifting by establishing a causal link between an occurrent event of willing and a subsequent event of arm lifting. It is counterintuitive to suppose that the causing intention disappears at the moment when the action begins. Instead,

> [f]or the raising of your arm to be an act, any willing must occur with the raising, at the same time. Any decision prior to the raising cannot be the cause of the raising because of the possibility of change of mind or forgetfulness. [. . .] the cause must continue to have some existence as long does the effect. [. . .] Your deliberate raising of your arm must continue all the way until the arm is as fully raised as you want. The willing, such as it is – for us a dispositional cause of the arm rising – is a power that must be exercising until the arm is fully raised as you want. (Mumford and Anjum 2011, p. 206)

Occurring simultaneously, cause and effect are 'reunited'. Mumford and Anjum illustrate this also in the following example:

> You lift up a book [. . .] and the book being lifted (the effect) is simultaneous with you doing the lifting (the cause). As soon as you take it from the table, the effect is in existence as well: the book is elevated. And as soon as you stop holding the book, and set it down, the effect ceases too. (Mumford and Anjum 2011, p. 114)

What are the ontological commitments of views on which mental dispositions play an active role in bringing about their manifestations? Ryle bases his conception of mental dispositions on an analysis of our everyday use of mental concepts. He aims to "determine the logical geography of concepts" (Ryle 1949, p. 8). This makes it hard to identify the ontological commitments of his analysis of mental concepts. However, we can sketch some of the ontological implications of his approach. On Ryle's view, mental concepts do not refer to mental or physical events *within* the agent. He explicitly denies the reality of mental events or substances as proper correlates of mental ascriptions (Ryle 1949, p. 22). The proper subject of mental predicates is the entire person with her specific way of acting. The bearers of mental dispositions are, on this conception, persons who act in characteristic ways:

> 'Know' is a capacity verb, and a capacity verb of that special sort that is used for signifying that the person described can bring things off, or get things right. (Ryle 1949, 117)

This seems also to be the case according to the dispositionalist view of Mumford and Anjum. Exercising the power of willing seems to presuppose an agent subject which endures as long as she is exercising the power of willing. According to Mumford and Anjum, such powerful processes are dynamic and "cannot be broken down into a string of changeless parts".[2] For this reason, they cannot be exercised by numerically different agents. In this case, each agent or temporal agent-part would exercise its own part of the willing-process and this would be the end of the dynamic unity of the exercising process (Mumford and Anjum 2011, p. 116).

Assuming that mental dispositions require endurers as their bearers, the further question arises whether this dispositionalist approach is compatible with the stronger ontological claim of enduring *powerful* agents. Can we assume that there are enduring agents who bring about their actions on the basis of specific dispositions? Mumford and Anjum claim that there is no strict necessity between a (dispositional) cause and its effect:

> Dispositionality is [. . .] not something as strong as a tie that binds things together inseparably. An effect does not always follow its typical cause. [. . .] It is, rather, that of something that disposes toward its effect, where the modality of dispositionality is *sui generis*, of its own kind, and certainly not reducible to pure necessity or pure contingency. It is something in between. (Mumford and Anjum 2011, p. 175; emphasis in the original)[3]

---

[2]See also Mumford (2009), p. 230: "[. . .] those [dynamic] properties that are essentially changes [. . .] thus cannot be instantiated at an instant. [. . .] The perdurantist cannot permit such essentially dynamic properties as real and irreducible."

[3]See also p. 176: "[. . .] to be disposed toward an effect or manifestation does not imply that such an effect or manifestation is necessitated."

If dispositions do not enforce their effects, one might wonder whether this leaves room for an agent to bring about her actions on the basis of her dispositions. Are mental dispositions open to different manifestations, and is it up to the agent to bring particular manifestations about? Answering this question requires, once more, a detailed analysis of the relationship between the mental power and its manifestation. Mumford and Anjum take agency to be the paradigmatic case of the special modality typical of dispositional causation. According to them, the "dispositional modality is the modality that we all know through our experience, as causal agents and patients. In this respect, it is the modal value best known to us" (Mumford and Anjum 2011, p. 176). What is the relationship, in the case of agency, between mental powers and their manifestations?

Mumford and Anjum describe a person's willing to lift her arm as a case where the person, as the agent, can have direct access (through proprioception) to her mental powers and their manifestations (Mumford and Anjum 2011, pp. 203ff). At first glance, it seems that the manifestation of a mental disposition is the corresponding behaviour. On this interpretation, the manifestation of the willing-power is the raising of the arm. However, a closer look at the Mumford-Anjum view of the manifestation of the willing-disposition reveals a more complicated picture: According to Mumford and Anjum, powers manifest themselves by being exercised. Exercising the will-power of raising one's arm is not necessarily the same as performing the act of raising one's arm. Mumford and Anjum describe the exercise of the willing as a process to which we have proprioceptive access: one can feel the effort through which a volition is exercised (Mumford and Anjum 2011, pp. 207ff.). Moreover, one can exercise the power of willing and yet fail to perform the action (Mumford and Anjum 2011, p. 212). We must thus distinguish between the effect, which is the raising of one's arm, and the manifestation of the willing-power, which is the execution of the volition and the effort to raise the arm.

This specification meets the requirements of George Molnar's distinction between the *manifestation* and the *effect* of a disposition. According to Molnar, each disposition or power is essentially connected with one specific manifestation: "Manifestations are isomorphic with powers because each power gets its identity from its manifestation" (Molnar 2003, p. 195). Effects, on the other hand, are *polygenic*: different powers, or rather the manifestations of these powers, contribute to a specific effect. If we apply this distinction to the case of willing to raise one's arm, there are different powers that contribute to the raising of the arm. Exercising the volition to raise the arm – the manifestation of the willing-power – is not enough; if I exercise too little power, or if my muscles fail to work properly, the effect will not occur.

According to Mumford and Anjum's dispositionalist model of mental powers, then, we must consider two different levels. The first level consists in the relation between mental powers and their manifestations, the second level consists in the relation between the manifesting and non-manifesting powers and the resulting effects. At the first level, Mumford and Anjum do not seem to assume powerful agents as the driving force of mental causation. Despite the fact that agency is believed to be a case in which we can perceive causation directly, there is no word about a subject who is a powerful agent and who would actively bring about the manifestation of the mental power of willing. Instead, it seems that the willing-power manifests itself as long as nothing interferes.

Let us turn to the second level of power-causation, the contingent relationship between different manifesting and non-manifesting powers and the resulting effects. What is the situation here? The effect – in our case, the arm's being lifted – is not connected in an a priori way with the manifestation of a specific power. The manifestation of the willing-power can occur without the effect. I can will to lift my arm without my arm's being lifted – for example, when there is an external obstruction so that the threshold required for the effect to occur is not reached. We thus come to see that on the second level too, the effect occurs with necessity if all of the required powers manifest themselves and if no other powers interfere. Nevertheless, Mumford and Anjum insist on the anti-necessitarian character of dispositional causation. Whatever we have to think about this, the thing that matters in the present context is that the assumed contingent relationship between the dispositional cause and its effect does not automatically open the field for *an agent* to bring about the effect. On the contrary, it seems that mental powers are on a par with natural powers. According to Mumford, "we have natural necessities without commitment to general laws ranging over classes of events. Instead of general necessities, natural necessity occurs at the level of the particular" (Mumford 1998, p. 221).

On the threshold model of causation, the agent, as the final determining causal factor, seems to disappear. Mental dispositions like willing are taken to be causally effective in combination with other powers. Even the manifestation of a power is co-determined by the interfering or enhancing manifestations of other powers. In this causal network of powers, there seems to be no space left for the agent to determine the outcome of the decision process.[4]

4.4 Agents in a World of Powers

Let me conclude with some final reflections on the possibility of agents in a world of powers. I have argued that there is no place for agents and agent causation in a world in which all effects are caused by dispositions. On the other hand, when it comes to the intuitive justification of the presumed distinctive modality of dispositional causation, Mumford and Anjum, as we have seen, appeal exactly to our self-experience as agents. Our self-experience as agents, however, comprises not only the proprioceptive awareness of the effort we must invest in order to exercise a

[4]Tim O'Connor discusses a similar problem when he criticises a probabilistic causal indeterminist account of human freedom: "Given the presence of desires and intentions of varying strength, making certain outcomes more likely than others, the agent possesses no further power to determine which outcome in fact is brought about. The determination is a product of the propensities of the agent's states, and the agent doesn't seem to directly control which propensity will 'fire'" (O'Connor 2009, p. 211).

power and perform the action. It also includes the basic intuition that we as agents bring about the action. The agent performs her actions through powers; but powers cannot replace the agent in her performing the action. Assuming enduring agents as bearers of mental dispositions makes sense only if those agents play an active causal role in the production of actions. Agents must be causally responsible for the manifestation of mental powers. As Mumford and Anjum themselves put it in a more recent article, agents "must cause their own behaviour through the exercise of their real powers" (Mumford and Anjum 2015, p. 144).

Is it possible to introduce real agency into Mumford and Anjum's threshold account of power-causation? As we saw above, Mumford and Anjum regard willing as the paradigmatic mental power. Since the power of willing is intrinsically connected with only one specific manifestation, namely the exercise of the willing-power, it seems that the willing-power exercises itself when nothing interferes. We need no agent to do the exercising. The picture changes, however, if we consider other mental powers or dispositions, like knowledge, beliefs, reasons, virtues, desires, and mental capacities. None of these types of mental dispositions is essentially connected with only one specific manifestation; on the contrary, they enable different manifestations.

In the second Chapter of the Book IX of *Metaphysics*, Aristotle distinguishes between rational and non-rational powers. The non-rational powers are directed at only one type of realisation; as Aristotle puts it, they produce "one effect; e.g., the hot is capable only of heating" (Aristotle 1928, *Metaphysics* IX.2, 1046b 6–7). The rational powers, by contrast, are – as Aristotle puts it – "capable of contrary effects" (Aristotle 1928, *Metaphysics* IX.2, 1046b 5). Aristotle discusses the example of medical knowledge, which is a power for producing health but also a power for causing disease. Although it proves problematic to equate rational powers with mental dispositions, I would like to go even a step further than Aristotle with respect to the potential manifestations of mental dispositions. Mental powers – as O'Connor puts it – are in fact "multivalent, capable of being exercised toward any of a plurality of options" (O'Connor 2009, p. 196), while physical powers are (normally) directed toward only one manifestation. This openness of mental dispositions for various – including opposite – manifestations requires a causally active agent.

O'Connor defines agent causation as "the power to directly determine which of several causal possibilities is realized on a given occasion" (O'Connor 2009, p. 211). Accordingly, he distinguishes between event-causal powers and agent-causal powers. Event-causal powers are "tendencies towards effects, i.e. the powers themselves are disposed to produce effects". Agent-causal powers, on the other hand, "confer a capacity upon agents to produce effects, i.e. the power is not disposed to produce anything, it merely confers on its possessor a generic disposition to act" (O'Connor 2011, p. 319). As such, agent-causal powers do not necessitate their effects. Reasons and other mental dispositions are, on O'Connor's view, probabilistically structured capacities. They confer a tendency to act in a specific way.

Only those mental dispositions are agent-causal powers that are exercised by an agent. But when are mental powers just event-causal powers, which trigger their effect if nothing intervenes, and when do they become agent-causal powers? A plausible answer to this question assigns a crucial role to self-awareness. O'Connor (2009, p. 199) suggests that a completely unconscious causal influence of cognitive preferences might diminish the agent's freedom to some extent. If reasons and other mental powers are completely unconscious, their influence on decisions cannot be part of the agent's considerations; they are beyond agent-causal control. It seems obvious that, in order to make a choice on the basis of reasons and other mental powers, the agent must have further cognitive abilities: she must in some way be able to represent these mental powers or – as Aquinas and Aristotle put it – she must have knowledge of them. Only if agents are able to represent their preferences, these preferences become agent-causal preferences; only in this case agents themselves – and not their internal powers – bring something about. More recently, also Mumford and Anjum have come to acknowledge an agent's need for a "higher-order power – a self-reflective one – to think about one's first-order powers (Mumford and Anjum 2015, p. 148). They emphasise that "we are free when we are able to exercise our causal will powers that can be regulated by higher-order powers of self-reflection" (ibid.).

What is it that agents bring about when exercising their agent-causal powers? O'Connor assumes that the agent who acts on the basis of her agent-causal powers brings about an act of will, not the action itself (O'Connor 2011, p. 314). An agent-causal power "confers upon an agent a power to cause a certain type of event within the agent: the coming to be of a state of intention to carry out some act, thereby resolving a state of uncertainty about which action to undertake" (O'Connor 2009, p. 195). In this respect, O'Connor's view is at risk of lapsing into the traditional account of mental causation, on which the agent brings about mental events, which then temporally precede the actions as mental causes. In accordance with the above-presented picture of classical philosophy of mind as to be found in Alvin Goldman, occurrent mental events, rather than agents, would be held to do the causal work in producing actions. Moreover, the exercise of willing itself would be beyond the agent's control.

Contra O'Connor, I contend that by exercising our mental dispositions we do not bring about an act of will but the action itself. Willing, on the other hand, is not a power which can or cannot be exercised. Instead, willing is inseparable from the action itself and, hence, (to put it in Mumford and Anjum's words) not something preceding the action but occurring simultaneously with it. For this reason, Mumford and Anjum's portrayal of willing as a paradigmatic mental disposition appears to be misleading. What the agent can exercise are mental dispositions that confer on its possessor a disposition to act; but an agent cannot in the same way exercise her willing of the action.

Let us summarise: All conceptions of mental dispositions discussed in this chapter require enduring bearers. There are, however, different views as to whether these bearers are also agents. In Goldman's higher order conception of mental dispositions, all the causation is done by mental occurrences or events. The conception of mental powers in the tradition of Gilbert Ryle, in which Mumford and Anjum stand, offers a more complicated picture. But it is open for agent causation only if we weaken Mumford and Anjum's threshold account of mental powers. An agent can play her causal role only if mental dispositions are open for different manifestations. This requires higher-order powers like self-reflection. By exercising our mental dispositions, we bring about actions. Willing is nothing else but the intentional bringing about of an action. The bringing about of the action is intentional because the agent performs the act by exercising her mental dispositions.

Acknowledgements I am grateful to the editor of this volume, Anne Sophie Meincke, for substantive editing as well as to Katherine Dormandy for helpful comments on earlier drafts of this chapter.

References

Aristotle. 1928. *Metaphysics*. In *The Works of Aristotle, Vol. VIII*, ed. W. D. Ross. Oxford: Clarendon Press.

Baker, Lynne Rudder. 2013. *Naturalism and the First-Person Perspective*. Oxford: Oxford University Press.

Goldman, Alvin I. 1970. *A Theory of Human Action*. Englewood Cliffs: Prentice-Hall.

Henderson, D., and T. Horgan. 2009. Epistemic Virtues and Cognitive Dispositions. In *Debating Dispositions – Issues in Metaphysics, Epistemology and Philosophy of Mind*, ed. G. Damschen, R. Schnepf, and K. Stüber, 296–319. Berlin: de Gruyter.

Lowe, E. Jonathan. 2010. On the Individuation of Powers. In *The Metaphysics of Powers*, ed. A. Marmodoro, 8–26. New York: Routledge.

Molnar, George. 2003. In *Powers – A Study in Metaphysics*, ed. S. Mumford. Oxford: Oxford University Press.

Mumford, Stephen. 1998. *Dispositions*. Oxford: Clarendon Press.

———. 2009. Powers and Persistence. In *Unity and Time in Metaphysics*, ed. L. Honnefelder, E. Runggaldier, and B. Schick, 223–236. Berlin: de Gruyter.

Mumford, Stephen, and Rani Lill Anjum. 2010. A Powerful Theory of Causation. In *The Metaphysics of Powers*, ed. A. Marmodoro, 143–159. New York: Routledge.

———. 2011. *Getting Causes from Powers*. Oxford: Oxford University Press.

———. 2015. Powers, Non-Consent and Freedom. *Philosophy and Phenomenological Research* 91 (1): 136–152.

O'Connor, Timothy. 2009. Agent-Causal Power. In *Dispositions as Causes*, ed. T. Handfield, 189–214. Oxford: Clarendon Press.

———. 2011. Agent-Causal Theories of Freedom. In *The Oxford Handbook of Free Will*, ed. R. Kane, 2nd ed., 309–328. Oxford: Oxford University Press.

Ryle, Gilbert. 1949. *The Concept of Mind*. London: Hutchinson.

Chapter 5
Powers, Activity and Interaction

Anna Marmodoro

Abstract Power Structuralism is a powers-only ontology, built on what is known as the Eleatic Principle: only what is causally powerful is real. In this chapter I concentrate on three distinctive tenets of Power Structuralism. I show how there is actuality at the bedrock of this ontology, without there being categorical (i.e., causally inert) properties. I distinguish between two types of power, and consequently two types of manifestation. The modality of power manifestations is conditional necessity. Causal interactions are explained in terms of a special type of monadic property – namely, directed power tropes, and their manifestation partners. Such powers are ontologically interdependent with their manifestation partners. I argue that the nature of causal interactions is relational; and bipolar.

Keywords Powers · Manifestation · Actuality · Causation · Relations and relatives

5.1 Introduction

Power Structuralism is a powers-only ontology built on what is known as the Eleatic Principle (Armstrong 1978, Vol. 2, p. 5): only what is causally powerful is real. In this chapter I concentrate on three distinctive tenets of Power Structuralism. Power Structuralism holds that the world is made of powers only, with no inert categorical properties or inert matter, and that a power exists either in potentiality or as manifesting (i.e. being exercised). Powers in this ontology are intra-structured and inter-structured, that is, internally structured and also composed into structures with

A. Marmodoro (✉)
Department of Philosophy, Durham University, Durham, UK

Corpus Christi College, University of Oxford, Oxford, UK
e-mail: anna.marmodoro@durham.ac.uk

A. S. Meincke (ed.), *Dispositionalism*, Synthese Library 417,
https://doi.org/10.1007/978-3-030-28722-1_5

other powers.[1] Here I show, first, how there is actuality at the bedrock of this ontology, without there being categorical (i.e., causally inert) properties. Second, I distinguish between two types of power, and consequently two types of manifestation. The modality of power manifestations is conditional necessity, that is, when in appropriate circumstances, powers manifest necessarily. Third, causal interactions are explained in terms of an ontology of a special type of monadic property, namely, directed power tropes, and their manifestation partners. Such powers are ontologically interdependent with their manifestation partners. I argue that the nature of causal interactions is relational; and bipolar.

The concern that ontologies admitting only pure powers (with no categorical, inert base) often give rise to is that they build the world out of pure potentiality. However, this concern rests upon the assumption that actuality is inertness, which I take to be a misconception. Positing inert categorical properties or any type of inert particulars does not answer or even address the question of actuality in the world. The answer to the quest for actuality, I hold, is to recognise that some powers are directly activated. Actuality is activity in my system, and since interaction is accidental (it may or may not happen), the activity actualising the world cannot be interactive; there has to be a different type of manifestation of powers. The intuition I am developing into a power ontology here is that change is not only change of another; it is also self-change. Change of another involves power interaction, whereas self-change involves engagement in a (non-interactive) activity. I present in what follows my arguments for these claims.

5.2 The 'Ghostliness' of Powers

In everyday life we often arrange (instances of physical) powers in some order or other, to perform certain functions; for example, complex power arrangements are common in factory production lines, in laboratories, in kitchens, and in every mechanism we build, from the large to the small. We arrange powers by ordering material objects in appropriate structures. Yet, how are powers arranged when the building blocks for the arrangements are not objects? If the most elementary constituents of the world are not material entities, but rather, in accordance with the Eleatic Principle, power tropes (or powers, for short) from which material objects and everything else in the universe are made up, how are these powers arranged? The question here is not (only) how powers compose and constitute macro-entities, such as material objects, bodies, processes, etc. It is (also) how powers compose and constitute elementary powers, such as an electron. And how they constitute the manifestations of elementary powers, such as the manifestation of (an electron's) electric charge as a wave, distributed throughout space. In general terms, do powers compose, and if they do, how do they compose?

[1]Power Structuralism is introduced in fuller details in Marmodoro (2017).

In my ontology, all the elementary entities in nature are built out of sparse properties,[2] which are powers. For example, electrons are elementary entities, and they are built out of their properties of mass, charge, and spin. Mass, charge, and spin are powers of an electron, which characterise it, and also enable it to do what it does; e.g., the mass that characterises the electron (co-)determines its variable weight in different contexts, and its charge (co-)determines its repelling and attractive forces. But mass, charge and spin do not, on their own, constitute the electron; there are numerous ways in which mass, charge and spin could come together into an entity, and most of these ways would not make up electrons (but rather entities which do not in fact exist in our world). The metaphysical difference between the mere compresence of mass, spin and charge, on the one hand, and the electron on the other is *structure*.[3]

Concerning the nature of powers, what is characteristic of them is that they can be exercised. For instance, the repulsive power of an electron is exercised when repelling another electron. But it need not be repelling all the time. At any one time, some powers in the world are being exercised while other powers are not being exercised, but are in potentiality. Herein lies a problem. What type of ontological item is a power trope in potentiality? More generally, what is potentiality? I hold that powers in potentiality are real. They are not empirically detectable in potentiality, because detection is causal interaction, which is the manifestation of the detected powers. Yet, powers in potentiality are indispensable for explaining the *possibilities* of interaction between elementary entities.

The undetectability of the indispensable might give rise to a worry. Is it possible that the most elementary stratum of reality consists of powers in potentiality? Could there even be a world of mere potentiality? That our world is a world of mere potentiality is the predicament of power ontologies that define manifestation as a relation between powers – from a potential power to another potential power that is the manifestation of the original one (e.g. Mumford and Anjum 2011). Such ontologies are committed to having in their system only potentiality, as David Armstrong famously objected. I refer collectively to the worries about the potentiality of powers as the "ghostliness" of powers. The solution to such worry cannot be to posit that in our world powers do manifest. Even if there exist powers in actuality, there is still a worry one may raise, that the ground level of reality is mere potentiality. In response to this concern, some metaphysicians introduce in their ontologies categorical properties (i.e., inert actualities) alongside powers, to guarantee that the bedrock of reality is actuality and not mere potentiality (as it would be,

[2] I use 'elementary entities' to refer to those entities which are fundamental in the ontology of the world. I use David Lewis' term 'sparse' for the elementary ontology of nature, meaning that 'sparse properties carve out the joints of nature on which the causal powers hinge'. See Schaffer (2004).

[3] *Pace* Aristotle, I argue that their difference is not a single structure, the substantial form, but structures of two different kinds. I argue for this point in my (2017).

if it comprised powers only).[4] Introducing categorical properties however raises a number of problems, already discussed in the literature.[5]

I turn now to examine in some more detail the ways in which the "ghostliness" of powers worry has been fleshed out in the literature. The first is about the possibility of change in a 'power only ontology', and is known in the literature as the *Always Packing Argument*.[6] David Armstrong (1997, p. 80) formulates it thus:

> Given purely dispositionalist accounts of properties, particulars would seem to be always re-packing their bags as they change properties, yet never taking a journey from potency to act.

The Always Packing Argument raises difficulties for views on which the activation of a power in potentiality is merely an instantaneous 'jump' to its manifestation, which is another power in potentiality. A theory of powers that did not allow powers, when activated, to exercise their powerfulness would be rather odd indeed – yet, this appears to be the position that most power ontologists are committed to, when claiming that the manifestation of a power is merely a new power in potentiality.[7] It is this conception of the power's manifestation that commits one to a network of powers in potentiality, where nothing ever is actual. This is the complaint that Armstrong's argument voices: that there is no actuality in the ontology. There is no journey from potency to act, not because nothing happens in a world of 'pure' powers; powers do in fact manifest; but the manifestation is an instantaneous transition to another power in potentiality. The exercise of powers on other powers does not get reified, i.e., it is not an item in the ontology, on such views. Change is not defined in terms of potency and act, but only in terms of potency to potency.

The second concern, about the possibility of there being things in a world of mere potentiality, is known as the *Domino Argument* and was put forward by John Heil. He writes:

> Despite its appeal in some quarters, many philosophers have been struck by the thought that a properties-as-powers view leads to a debilitating regress. Suppose *A*s are nothing more than powers to produce *B*s, *B*s are nothing more than powers to produce *C*s, *C*s are nothing more than powers to produce *D*s . . . and so on for every concrete spatio-temporal thing. How is this supposed to work? Imagine a row of dominos arranged so that, when the first domino topples, it topples the second, which topples the third, and so on. Now imagine that *all there is* to the first domino is a power to topple the second domino, and *all there is* to the second domino is a power to be toppled and a power to topple the third domino, and so on. If all there is to a domino is a power to topple or be toppled by an adjacent domino, *nothing happens: no domino topples because there is nothing—no thing—to topple*. (2003, p. 98, my emphasis)

---

[4]See, e.g., Armstrong (1997), p. 79.

[5]See Bird (2007), ch. 4, for an insightful discussion. The main concern with categorical properties is their inertness, which cuts them off our cognitive reach, since they cannot interact with us in any way.

[6]The expression "always packing, never travelling" is first used by Molnar (2003, p. 173).

[7]See, e.g., Mumford and Anjum: "The manifestation of a power will [. . .] be itself a *further* power or cluster of powers" (2011, p. 5, my emphasis).

We saw that the Always Packing Argument assumes that the world is a sequence of powers in potentiality getting constantly replaced, upon being manifested, by other powers in potentiality; there is nothing but powers in potentiality in the world. The Domino Argument does not question how a power will take the journey from potency to act, if all there is to a thing or to the ontology is potency. Rather, it questions how a power can take the journey from potency to act *if all there is to the thing a power will act on is potentiality*. This is a concern, not about the fact that powers never take a journey from potentiality to actuality, but about their journey having no destination, as it were. In other words, the question is whether there is an object *at all* when all there is to its constitution are pure powers in potentiality.

Are mere collections of powers in potentiality *no thing* at all? In view of the considerations about categorical properties above, I consider that there is no metaphysical impediment to the existence of things constituted of pure powers in potentiality. On the other hand, such a circumstance could not arise in our world. As a matter of fact, things in our world are not combinations of powers in potentiality only. Fundamental powers of elementary particles are continuously manifesting in their environment, given the presence of the ambient gravitational force in the universe, but also of the other fundamental forces. In complex objects, powers are continuously manifesting in the presence of other powers in the same object; for instance, the gravitational and electromagnetic powers of some of the domino's physical parts manifest in the presence of such powers of other parts of the domino, etc.[8] Powers such as the power to topple presuppose structural constitutional complexity of the object that possesses them – size, shape, weight, hardness, etc. Many of these powers of a domino will be constantly activated, manifesting in the presence of the other powers of the domino, even if the power to topple is not manifested. In short, there is a domino there to be toppled.

The alternative power ontology I want to argue for, Power Structuralism, is not susceptible either to the charge that its worlds would be worlds of mere potentiality, with no change, or the charge that it lacks actuality. The reasons have to do with how Power Structuralism accounts for change and, hence, the manifestation of powers. Operating within the scope of the Eleatic Principle, I want to introduce a distinction between what we may call *intransitive* change and *transitive* change: namely between *changing*, and *changing another*. As applied to power ontology, the distinction is as follows:

Intransitive change is activity that is directed towards itself, e.g. the spin of a particle; the movement of a photon in empty space; etc.

Transitive change is activity that is directed towards something other, e.g. the magnetic attraction of another magnetic charge; the fundamental interactions between the elementary physical systems; etc.

[8]The four fundamental forces, the week nuclear, the strong nuclear, the electromagnetic and the gravitational forces, are not additional powers for physicists. Rather, they are considered four fundamental interactions, explained as the emission and absorption of elementary particles. So, following my description above, the four fundamental interactions are emissions and absorptions of combinations of fundamental powers constituting elementary particles.

I consider the distinction between intransitive and transitive change to be fundamental in nature, and fully recognised in physics. I will accordingly distinguish between 'intransitive powers' and 'transitive' or 'interactive powers'.

Intransitive powers are exercised in their characteristic activities. For instance, the power of spin of an electron is exercised in the activity physicists describe as the spin of the electron. Some intransitive powers are *always exercising*. For instance, the power for electric charge of an electron is always exercising as a wave that generates an electric field. There are no powers of electric charges in potentiality, i.e., not manifesting in wave-form activity. Some intransitive powers on the other hand are not always exercising. For instance, an electrostatic charge has potentially the power to generate a magnetic field; but it generates a magnetic field only when it is in motion. Hence, a motionless electrostatic charge possesses a power of magnetism in potentiality only. Interestingly, not all the sparse powers in nature are intransitive powers. Some of the sparse powers that physics discovers at the bedrock of reality, in terms of which the elementary particles are defined, are transitive or interactive powers. For instance, mass results from the interaction of fields, e.g. between a photon and the Higgs field. So mass results from the interaction of sparse powers.

Having introduced intransitive powers, I want to return to the objection to pure power ontologies on account of their allowing for worlds of mere potentiality. In the ontology of Power Structuralism, I define:

> The *real* as powerfulness (as *per* the Eleatic Principle);[9]
> The *actual* as manifesting-powerfulness, namely the activated subclass of the real.

It follows that everything in the ontology of Power Structuralism is real, *qua* powerful; and some of what is real is actual (what is manifesting is actual); while some is potential (*qua* unmanifested). Hence, the omnitemporally manifesting activities, such as the Higgs field, are the actuality at the bedrock of the universe. Such activities, as a charge manifesting as an electric field, comprise *elementary activity* and hence, *elementary actuality* in the universe. Elementary activities ground further powers in potentiality, such as magnetic powers. They also ground, as we will see, compositions of powers. There is therefore no need to posit categorical properties to avoid worlds of mere potentiality; occurrent activity suffices for actuality. Our world is a dynamic world of occurrent activity and potential activity.

Turning now to the *interactive powers*: as we have seen, activities, such as an electric field, have intransitive powers of their own; for instance, the power of a magnetic field that electric fields possess potentially, which is activated when they are in movement. But as indicated above, activities do not have only intransitive powers of their own. They also possess interactive powers, through which they interact with other activities, e.g. in the case of the interaction of two electric charges. Interactive powers are exercising when the activities that possess them (e.g. photons) interact through them. When not exercising in an interaction, the interactive powers are in potentiality. When interactive powers interact, their manifestation lasts while

[9]More explicitly, it is powerfulness that is real rather than any carrier of it.

they are interacting; their manifestation is simultaneous, in a mutual exercise on each other. Interactions are causal activities that result in *changes* in the interacting activities. The changes brought about in mutually interacting elementary activities are quantitative, and measurable in physics. Physics quantifies changes of mutually interacting elementary forces in terms of exchanges of virtual force carriers between them. What was described in classical physics as the action of a force field generated by one particle on the other, is now described in quantum physics as the exchange of virtual force carrier particles between them.[10]

Causation is involved not only in interactions; as we understand the term from the Eleatic Principle, causal activity is the manifestation of any kind of power: intransitive powers, or transitive, i.e. interactive, ones. Powers interact with one another when in appropriate circumstances, e.g., when in contact with one another, as in the case of crisscrossing fields of electric charges. Each power manifests, doing to the other power what it is its nature to do; thus negative electric charges repel one another; the activity of their interaction is to repel each other. On a secondary conception of 'causation' introduced here, intransitive powers are also causal, in so far as they bring about occurrent change – activity. Activity is change, not of another, but of the activated entity. Our 'working' conception of causation, however, that to which we appeal most often, is that of interaction of powers.

Every interactive power has partner powers it can interact with.[11] For example, repelling does not happen on its own, or in response to any type of power. When, e.g., a charge repels, it repels another charge; or more generally, coupled elementary fields can interact between them through their interactive powers. Partner powers are *ontologically interdependent* with one another. Their definitions establish their interdependence. For instance, the definition of the magnetic moment of a magnet states that it determines the turning-force the magnet will experience when in an external magnetic field, e.g., what a compass needle experiences in the Earth's magnetic field. Here the magnetic moment of a magnet is defined in terms of how the magnet would interact with an external magnetic field, which establishes their ontological interdependence.

[10]See, e.g., "Virtual particles are viewed as the quanta that describe fields of the basic force interactions, which cannot be described in terms of real particles" (http://dictionnaire.sensagent.leparisien.fr/VIRTUAL%20PARTICLE/en-en/, last accessed on 8 May 2019).

[11]By contrast, intransitive powers do not have partner powers. E.g., a wave field manifests without interacting. (Some intransitive powers need special conditions to obtain so as to manifest, as in the case of generating a magnetic field from an electric one; but these special conditions are not the same as partner powers.)

5.3 Relationality

Interactive powers are not relations, but are *relational*. They are not polyadic relations. Their directedness does not constitute any type of physical bridge between themselves and the power they are directed towards. Interactive powers are *monadic properties*, which are directed towards other properties. They are *relational* in virtue of their directedness. By this I mean that their directedness translates into relational behaviour: it regulates a mutual manifestation between them. Thus partner powers are monadic, but behave in concert with one another when manifesting.[12] The modal behaviour of partner powers that interact by mutually manifesting stems from their natures – their essences – as powers[13]; e.g., electric charges repel others in appropriate conditions. As we have seen, their directedness is grounded on their ontological dependence on each other. So, repelling powers are, by their definition, triggered into repelling behaviour by the type of power they can repel. They are directed towards such a type of power even when in potentiality, since this is their nature; for instance, an electron possesses the power to repel another electron, whether there is another electron in its vicinity or not. (Note that being interdependent is not sufficient to make partner powers directed towards each other; for instance, a car is not directed towards petrol, even if it is dependent on it to move. There are further conditions that obtain in the case of partner powers; but I will not here examine the full range of conditions for being directed.)[14]

It is important to appreciate that the account of the fundamental interactions among elementary forces that physics offers in terms of the exchange of virtual particles (force carriers) does not supersede the need for, and the role of, the essence of powers for their interactions. The exchange of virtual particles is a description of how the forces interact between them. But what needs further explanation is the fact that one force affects the other. Regardless of *how* they affect one another, whether through virtual particles, or plane waves, or wave disturbances, etc., the question is, *why* do they affect each other at all? For example, electric charges do interact with one another; they don't pass each other by. This capacity to interact with one another that characterises electric charges, as opposed to other types of radiation, needs to be explained and is not explained by the exchange of virtual particles. The mechanism of the virtual particles presupposes the capacity to interact, to affect one another. It is the nature of the powers that determines their 'sensitivity' to each other, and dictates their behaviour towards each other.

---

[12]The point holds of two or more partner powers. I talk of two only for simplicity.

[13]I follow Kit Fine (1994) who has developed and argued for the Aristotelian position of modal primacy of essence over necessity.

[14]Compare, for instance, Schaffer (2004), p. 214: "causation has a counterfactual aspect, involving a comparative notion of difference making. I leave open whether causation is a purely counterfactual affair, or whether it involves some hybrid of counterfactuals and physical connections". See also Schaffer (2000). What Schaffer calls 'difference making' is comparable to the directedness of powers as an aspect of the phenomenon of causation; for instance, the pot became hot due to the fire, and the fire was directed towards heating another (e.g. the pot).

5.4 The Bipolarity of Causation

What type of event is an interaction between two (or more) powers? For instance, when two charges repel each other, what is going on between them? It is a single interaction, which consists of two reciprocated repelling actions. My claim is that this is the pattern of interaction of partner powers. There is a single interaction between, say, two powers, which is characterised by two types of action, each characteristic of each of the two interacting powers. In the present example, the two types of action are mutual repellings of each other in *opposite* directions. In other kinds of case, the two (or more) types comprising the interaction might be mutual gravitational attraction in opposite directions; or emitting particles in opposite directions, etc. Generally, the interaction between partner powers comprises two activities, each of which has the opposite direction from the other. In metaphorical terms, the partner powers interact like two boxers actively moving against one another, rather than like two synchronised swimmers. If one partner gives, the other partner takes. And vice versa.

On the widely accepted view that properties characterise through their natures the subjects to which they belong, how should the interaction between two partner powers characterise each of their subjects or bearers? In the last example, does it make its bearers givers or takers? The short answer is that each bearer is characterised by the manifestation of that among its powers which constitutes the interaction. One component of the interaction makes one of the partners a giver, and the other a taker; and the other component of the interaction makes the taker a giver, and the giver a taker. It follows that the interaction between partner powers is an activity that is *bipolar* in nature, exhibiting the asymmetry of the directedness of each of the partner powers, which are directed towards each other. Their interaction comprises opposed sub-activities that are mutually sustained by exercising against each other.

Although the manifestation of (interactive) partner powers is bipolar, it should not be thought that this is the case also with the manifestation of intransitive powers. Thus, although electrons repel each other, on their own they are active by being electric fields.

5.5 The Modality of Powers

The essential dependence of (interactive) powers on their partner powers for manifesting under appropriate conditions establishes an order of *conditional necessity* in the operation of nature. Such powers operate under conditional necessity, as we understand their behaviour in the physical sciences. That is, the behaviour of macro-objects in natural phenomena is deterministic, which, for the powers involved in them, means that when the conditions, as *per* the natures of the powers, are right, the powers manifest.

At the micro-level on the other hand, we find a different type of determinism. As Stephen Hawking and Leonard Mlodinow explain,

> Quantum physics might seem to undermine the idea that nature is governed by laws, but that is not the case. Instead it leads us to accept a new form of determinism: Given the state of a system at some time, the laws of nature determine the probabilities of various futures and pasts rather than determining the future and past with certainty. (2010, p. 72)

This suggests that in place of classical determinism, which allowed for precise prediction of future physical values, we have *determined probabilities* of future physical values. Determined probabilities should not be understood as advocating that powers manifest probabilistically. Again, leaning on Hawking's (1999) words in his lecture 'Does God Play Dice?':

> If you know the wave function at one time, then its values at other times are determined by what is called the Schroedinger equation. Thus one still has a kind of determinism, but it is not the sort that Laplace envisaged. Instead of being able to predict the positions and speeds of particles, all we can predict is the wave function. This means that we can predict just half what we could, according to the classical 19th century view. Although quantum mechanics leads to uncertainty, when we try to predict both the position and the speed, it still allows us to predict, with certainty, one combination of position and speed.

Physics studies and measures the micro-level phenomena that are governed by determined probabilities. But this does not mean that conditional necessity does not apply to the microcosm; it does apply, not to the individual causal phenomena, but to probabilistic patterns of their combinations, as Hawking explains.

The question of chance and determinism is extremely complex and controversial. Building the ontology of Power Structuralism, I am guided by a consensus that is developing in the literature about the compatibility of determinism with chance, when determinism is moderated by probability. Eagle (2019) writes:

> The general technique is to argue that there are probability distributions over outcomes that can play the chance role, even in impeccably deterministic theories. Many of these philosophers[15] are sympathetic to reductionism about chance, which permits the compatibility of chance and determinism. For the fact that the entire history of a world supervenes on any moment of its history, as determinism states, apparently entails nothing one way or another about whether the best description of that entire history involves chances.

Probability distributions of outcomes are surface phenomena, which may be underpinned by any number and sort of mechanisms, including deterministic mechanisms. The challenge for the advocate of chance is to offer an account of the *mechanism of chance, which gives rise to a probability distribution*. What is difficult to comprehend is the assumption that there is such a mechanism, made merely on the basis of an observed probability distribution and the failure of local deterministic explanations.[16]

---

[15]Eagle (2019) refers to works by himself; Luke Glynn; Carl Hoefer; Jenann Ismael; Eliot Sober.

[16]As, for instance, in the case of Bell's Inequality Theorem.

5.6 Conclusions

All there is in the elementary world is activity and interaction among activities. Both activity and interaction are manifestations of powers. Some intransitive powers, for instance an electric charge, are always manifesting[17] as a matter of physical laws; while transitive powers, e.g. a charge's magnetic power, manifest only in appropriate circumstances, i.e., in the presence of partner powers.

All powers bring about change. But intransitive powers are not causal powers in the sense in which interactive powers are causal. The former bring about change in the form of their own activity; while the latter change another power and its bearer. Intransitive powers are not themselves interactive. But they *have* causal powers, namely, they have interactive powers through which they interact with other powers, such as the powers to repel and attract other charges.

Interactive powers have partner powers with which they mutually manifest. Partner powers are monadic properties, but exhibit relational behaviour by being directed towards each other, in virtue of their essences that are ontologically interdependent. Their manifestation circumstances vary according to their type; for instance, some manifest when activities are in 'contact' with one another, as in the case of crisscrossing fields of electric charges.[18] Their mutual manifestation is the relational dimension in their modal behaviour. Modally, it is governed by conditional necessity (although it is different aspects of mutual manifestations that are so governed at the macro-level and the micro-level phenomena). Finally, the nature of the interaction between two manifesting partner powers is bipolar, characterised by the natures of the two interacting powers respectively.

The exchange of virtual force carriers between the fundamental forces in nature, as assumed by latest research in the physical sciences, does not explain all aspects of change. On the one hand, there are changes happening in nature which are not exchanges, such as the movement of a photon. On the other, exchanges of virtual force carriers explain the quantitative aspects of change, but merely posit qualitative aspects of change (e.g. the formation of a particle as a result of an exchange of virtual particles). On the view I have developed, qualitative aspects of change would be explained either in terms of the nature of the powers in the phenomenon, or in terms of their structure.

In conclusion, in this chapter I have introduced an original ontology that comprises only instances of physical powers at its fundamental level. Powers are the building blocks of the universe and they are real, even when they are not active. There are transitive and intransitive powers: the latter are always active, whereas the former are activated in virtue of their interaction with other powers.

---

[17]Intransitive powers which are not always manifesting may come to manifest, not by interacting, but by the obtaining of appropriate enabling conditions.

[18]The concept of 'contact' here is generic, to be interpreted by physicists, e.g., as overlap of fields.

Acknowledgements The research leading to this publication benefitted from a fellowship at the Paris Institute for Advanced Studies (France), with the financial support of the French State, managed by the Agence Nationale de la Recherche, programme "Investissements d'avenir", (ANR-11-LABX-0027-01 Labex RFIEA+). Thanks are due to the editor of this volume, Anne Sophie Meincke, for helpful feedback on the penultimate version of the chapter.

References

Armstrong, David. 1978. *Universals and Scientific Realism*. Cambridge: Cambridge University Press.

———. 1997. *A World of States of Affairs*. Cambridge: Cambridge University Press.

Bird, Alexander. 2007. *Nature's Metaphysics: Laws and Properties*. Oxford: Oxford University Press.

Eagle, Antony. 2019. Chance versus Randomness. *The Stanford encyclopedia of philosophy* (Spring 2019 Edition), ed. Edward N. Zalta. https://plato.stanford.edu/archives/spr2019/entries/chance-randomness/. Last accessed on 10 May 2019.

Fine, Kit. 1994. Essence and Modality. *Philosophical Perspectives* 8: 1–16.

Hawking, Stephen. 1999. *Does God play Dice?* http://www.hawking.org.uk/lectures.html. Last accessed on 10 May 2019.

Hawking, Stephen, and Leonard Mlodinow. 2010. *Grand Design*. New York: Bantam Books.

Heil, John. 2003. *From an Ontological Point of View*. Oxford: Oxford University Press.

Marmodoro, Anna. 2017. Power Mereology: Structural Versus Substantial Powers. In *Philosophical and Scientific Perspectives on Downward Causation*, ed. M. P. Paoletti and F. Orilia, 110–127. Oxford: Routledge.

Molnar, George. 2003. In *Powers: A Study in Metaphysics*, ed. S. Mumford. Oxford: Oxford University Press.

Mumford, Stephen, and Rani L. Anjum. 2011. *Getting Causes from Powers*. Oxford: Oxford University Press.

Schaffer, Jonathan. 2000. Causation by Disconnection. *Philosophy of Science* 67 (2): 285–300.

———. 2004. Causes Need Not be Physically Connected to Their Effects: The Case for Negative Causation. In *Contemporary Debates in Philosophy of Science*, ed. C. Hitchcock, 197–216. Oxford: Blackwell Publishing.

Chapter 6
What Are Manifestations?

Neil E. Williams

Abstract There is widespread agreement among friends of irreducible powers that powers are properties, and intrinsic ones at that. But when it comes to *what* is produced by the exercise of powers, little has been said about what sort of ontological category these 'manifestations' belong to, and even less agreement is to be found. The default response seems to be that manifestations are events, but a few recent proposals treat manifestations as properties or processes. Though these proposals boast a number of virtues, I argue that any view that understands manifestations these ways has unsavoury consequences that outweigh any virtues on offer. Those unsavoury consequences come to light when we consider manifestations that go awry. In particular, I consider an attempted poisoning foiled by antidote, wherein the circumstances required for the power's triggering have been satisfied, but the anticipated causal chain is frustrated. One the one hand, it seems mistaken to say that the power has been manifested, as the characteristic manifestation—death—has been avoided. But on the other hand, it seems equally wrong to claim that the power was not manifested, given that the conditions required for the display of the power were met and the expected causal process began, despite not running to completion. I thus propose that we think of manifestations as being extremely short-lived, ideally as states of affairs. En route to that conclusion some important discoveries about powers-based causation are made apparent.

Keywords Dispositions · Powers · Manifestation · Causation · Ontology · Antidote · Event · State of Affairs · Property

---

N. E. Williams (✉)
Department of Philosophy, University at Buffalo, Buffalo, NY, USA
e-mail: new@buffalo.edu

A. S. Meincke (ed.), *Dispositionalism*, Synthese Library 417,
https://doi.org/10.1007/978-3-030-28722-1_6

6.1 Four Candidate Categories

A great deal of recent work has been dedicated to understanding the nature of irreducible causal powers and how they are exercised. There is widespread agreement among friends of irreducible causal powers that powers (or clusters of them) are properties, and intrinsic ones at that. And there is a generally shared impression that powers are exercised in concert with other powers in an egalitarian way for a product that is generated mutually. However, when it comes to *what* is produced by the exercise of powers, little has been said about what sort of ontological category these 'manifestations' belong to, and even less agreement is to be found. In fact, despite the large amount of attention paid to powers and how they bring about their manifestations, an embarrassingly small amount of that attention has been directed at the nature of the manifestations themselves. This chapter is an attempt to redress that issue.

Examples taken from the powers literature indicate that manifestations fall into one of the four following ontological categories: properties (fragility results in *being broken*); states of affairs[1] (flammability results in *something's being combusted*); events (solubility results in *dissolving*); or processes (gravity results in *acceleration*). If number of appearances in the literature is anything to go by, then the event category boasts an early lead, but I suspect that this has as much to do with the popularity of certain sorts of examples as it does with anything else (pass the solubility, please). The event category was also a common choice among those first to offer analyses of powers: to wit, Ryle (1949) and Carnap (1928, 1936–1937) both took manifestations to be events. But considering that neither was at all friendly to irreducible powers—both sought to have them quietly removed from the ontological scene—we had best not put too much stock in what they had to say either. And finally, though Choi's survey article suggests that the manifestation of solubility might "uncontroversially" be "the event of dissolving" (Choi 2018), Mumford and Anjum (2011) claim without reservation that powers "manifest themselves in properties" (p. 5), and Lowe (2006) asserts that the manifestation of brittleness is "essentially a process" (p. 170), so it looks like the question of which category manifestations belong to is far from settled. I will therefore treat it as an open question, and will assume that we can trust the examples at least to the extent that whatever the correct answer is, it is to be found within the four aforementioned options.

I will also assume—perhaps more contentiously—that manifestations belong to one—and only one—fixed category. Hence, in addition to assuming that manifestations belong within the categories listed above, I will assume that there is a *single* correct answer. Why might one resist this?

One reason you might object is because you think that manifestations belong to some category *other* than the four listed. Though a handful of other candidates are

[1]Throughout the chapter I follow Armstrong (1997) in understanding states of affairs as instances of particulars instantiating properties, not as abstract objects.

available (facts or substances, perhaps), I take the four options on the table to be the best and most promising, and those to which the extant literature most clearly points. The same response applies to those who might suggest that manifestations constitute a sui generis category: that is not the story the literature tells, nor the direction our intuitions would have us go.

A second (and more compelling) reason to reject the assumption that the correct answer is one of the four listed is that you think that no single category can do the job on its own. That is, despite agreeing that all powers are properties, one might nevertheless think that manifestations can vary in their ontological category. Consider the following quotation, taken from Mumford and Anjum (2011): "Depending on its partnerings, heat can produce expansion, explosion, melting, boiling, steam, fever, burning, fire, pleasure, pain, growth, life, death, and so on" (p. 35). To my mind, life is as good an example of a process as one can hope to find, and probably growth and melting are processes too. But boiling sounds like an event, as does death. Fever strikes me as more state-like (certainly if defined as registering a temperature at or above 100 degrees Fahrenheit), and steam and fire could easily be interpreted as properties. You might disagree with my brief rundown, but let us bracket those concerns for now. I do not yet care to engage in any serious quibbles about how any of these is best understood; all I want to point out is that a legitimate case might be made for denying that manifestations fall into a single category. Given that heat can produce this range of disparate manifestations, if nothing can be done to tame this motley crew, then perhaps we should be pluralists about the right ontological category for manifestations.

But I want to resist all temptations to pluralism. This case, and those like it, are somewhat compelling, but all they demand is that we have different ways to *describe* some portion of the world when and if a manifestation takes place.[2] And we can do that perfectly well with a single fixed category. If, for argument's sake, we took the right category to be that of events, then when a manifestation event occurred it would be open to many descriptions. Some of those would be within the event category (a boiling can be a violent boiling, a boiling over, a rolling boil, and so on), but that same event has as constituents various states and properties (locations of the molecules, chemical states, specific temperatures and pressures, and so on), and it is part of a process, if it is not one itself. Case in point, a single event makes true ever-so-many claims, where those claims might be about the event as a whole, some portion or aspect of it, or what you get if other events of the right sort succeed it, but none of this threatens our supposition that manifestations are events.

A final problem one might have with my assuming that there is a single category to which manifestations belong concerns the categories themselves. On any reasonable understanding of what it is to be a property, state of affairs, event, or process,

[2]A general lesson of metaphysics we all ought to heed is that one should never trust how we speak about things as offering serious evidence about the ontological facts. Language can sometimes offer a modicum of guidance—and occasionally is the best and only thing we have to go on—but we only should trust it as far as we can throw it.

one or another of these categories is going to crosscut another. One can provide clear enough examples of each, but there is simply no denying that some or all of these categories overlap. To wit, any process is also an event, is it not? Worries aside, it is my plan to stick with the low hanging fruit and treat these categories as if they are much more clearly delineated than they really are. I will do this by forcing them into a very rough ordering on the basis of paradigmatic temporal extent. They will still admit of overlap, but there will be non-overlapping regions of each. Hence, I treat properties as ways that substances can be and thus, in a slightly odd sense, instantaneous, whereas states of affairs can obtain both at instants and for longer. Paradigmatic events will be four-dimensional entities with a temporal extent, but they will typically be shorter-lived than processes, which I am supposing are protracted and structured, in part because they are composed of many states and events.[3] I appreciate that this ordering is imprecise and that there are straightforward counter-examples to even my rough divisions, but I will ask that you bear with me for the time being. If we are to make any progress at all on the question of which ontological category manifestations belong to, then we have to start somewhere. As we shall see soon enough, a precise answer to the question will prove difficult to find, but we will have come much closer to finding one if we can rule out certain candidates.

I will concede that I am hard pressed when it comes to providing strong positive reasons in favour of approaching the question as if the answer must be monistic. The standard meta-ontological virtues of simplicity and elegance breed a preference for ontological tidiness, and perhaps that is enough. Nevertheless, I believe that any prima facie pull in the opposite direction can be resisted.

With that in mind, it is time to show my hand. I think the best answer to the question is that manifestations are states of affairs. The result of a power's being exercised is that there is some state of the world. Typically this will be a matter of a substance's instantiating some property or properties. This is what I will be arguing for in what follows. However—and this will sound very odd indeed—I will not be upset if I fail to convince you of the same. I prefer to treat manifestations as states of affairs, but I am not against interpreting them in other ways. Not, that is, unless one insists on claiming that manifestations are entities of any significant temporal extent. I will consider my argument successful if I can convince you that manifestations—whatever one takes them to be—must be the sorts of things that are very short-lived, and in some cases, instantaneous. Hence my primary aim is to argue that whatever ontological category manifestations belong to, it had best sit well with their being incredibly brief. It is the need for brevity that underlies my preference for treating manifestations as states of affairs, but I am open to their being events, if only in the minimal sense that the instantiation of a property at a time or over a brief period of time is an event (but not in the way that a dissolving, a burning, or a shattering is an

[3]I am not ignoring the fact that events—especially on Jaegwon Kim's (1973, 1976) analysis, wherein they are constituted by the exemplification of a property at a time—can be instantaneous, or that they too can be broken up into smaller events, but we are sticking close to the paradigm cases. Where serious problems for this thinking arise, the matter will be discussed in detail.

event, as these are all too long-winded). My ordering of the categories is designed to reflect this division: properties, states, and events (if properly construed) reside at the briefer end of the temporal scale, and are therefore distinguished from the processes and paradigmatic events that occupy the other.

With that coarse structure in place, let us begin our task of finding the right category for manifestations.

6.2 Manifestations: The Basics and Beyond

Whether you consider yourself a close friend of irreducible causal powers or merely tolerant of them, you should have no trouble agreeing with the following facts about manifestations:

1. We say of a power that it is 'manifested' when it is exercised. A manifestation is the *display* of a power—it is the way in which a power presents itself. Manifestations then are what powers produce when enacted; they are the outcomes towards which powers are oriented. This is not to suggest that only their displays are real, or any other such super-empiricist Rylean rubbish, only that there is a distinction to be made between having a power and exercising a power. When powers go unmanifested it is the manifestations that are then unreal, not the powers that produce them.
2. Because powers are capable of lying undetected (when the conditions appropriate for their display do not obtain) we are fond of picking out powers by way of the manifestations they produce; when in fact they do. To be flammable is to have the power to be ignited; to be soluble is to have the power to go into solution; to be fragile is to have the power to break, and so on. There may be types of powers that have never been exercised, and consequently we lack names for them. This makes them no less real than does their ability to be manifested in ways other than those by which we identify them.
3. Though we speak of a single power and *its* manifestation, this is only part of the story. For most of the manifestations we are interested in, the manifestation is produced by many powers acting in concert.[4] When a power finds itself appropriately arranged with other powers, those powers *jointly* act to bring about what is their mutual manifestation. Producing manifestations is therefore a group effort. This does not make it incorrect to say of each of the powers involved that it is for that manifestation—no more than saying that the I-90 highway will get you to Cleveland; it does, but so will the 77, the 71, and the 80—it just omits some relevant information.

[4]There is a class of powers whose manifestations are not generated co-operatively like this. They are what I have dubbed 'unilateral powers' (Williams 2017), but as they do not pose any special problems for the present discussion, I shall leave them aside.

4. It is a similarly innocent omission to speak of *the* manifestation a given power can produce, as any single power is capable of producing many different manifestations with different partners. Powers are for a great many different manifestations; which ones get produced typically depends on the presence and arrangement of other powers (that is, which constellations obtain) (Williams (2011); see Lowe (2010) for a friendly difference of opinion).
5. Manifestations involve further powers. Whatever manifestations turn out to be, they have as constituents (or are, or result in—this has yet to be seen), more powers. Once the solubility of salt is manifested by its dissolving in water it is capable of being precipitated; broken glasses can be used as cutting tools; and so on (Martin 2008, p. 91).
6. Some powers are always manifested (such as gravitational force), whereas others are manifested only on the presence of other partners (flammability). Cases of the latter sort underlie talk of the 'conditions' in which a power is manifested, but it is mistaken to think that conditions are relevant to the exercising of all powers.

That gives us a starting point and some background in considering which category manifestations belong to.

6.3 A Near Fatal Poisoning

To help us on our way—and to get the cognitive juices flowing—I want to open the discussion by focusing on a specific sort of power, namely that of being a poison.[5] Though there are plenty of technical and more nuanced ways of understanding what it is for a substance to be a poison, for present purposes we will do fine sticking to the colloquial definition: for a substance to be a poison is for it to have the power to cause death when ingested. Now consider the following scenario:

> Two brothers, Angus and Malcolm, are hiking in Australia when Angus has the misfortune of being bitten by a highly venomous Tiger Snake. His foot swells at the location of the bite, and Angus starts to sweat as his breathing becomes laboured. Lucky for him, Malcolm is no stranger to snake bites, so quickly splints and binds the area, thereby limiting the flow of venom. Nevertheless, a small amount of the potent neurotoxin in the venom begins to work its way through Angus's system, and he starts finding it harder and harder to move. Malcolm resorts to carrying Angus, and in due time arrives at a local emergency ward, where an anti-venom is administered. Angus recovers, and has little to show for his encounter beyond a minor wound and an exciting tale.

Even rare visitors to the powers literature should find this sort of case familiar. It is, after all, exactly the sort of eponymous 'antidote' case that Alexander Bird (1998) raises in response to David Lewis's revised form of the conditional analysis (Lewis

[5] I am well aware of the pitfalls we might encounter by directing the discussion in this way and the worries arising from the case of poisons in particular; I attempt to assuage those concerns in due course.

1997). But my raising of the case here has nothing to do with that issue.[6] The reason that the case is relevant to our present concern is similar to what we find in the medical field: it is that lesions can be informative about what constitutes health. Mutatis mutandis, we can learn a lot about manifestations from the poisoning scenario precisely because it goes awry. The antidote throws a proverbial spanner in the manifestational works, and once we have figured out how to deal with that, we will have gone a long way towards answering the question at hand.

So what has gone 'wrong' here, and what should we say about it? The sense in which the poisoning case has gone wrong concerns its failure to have resulted in Angus's death. The poison was ingested, but counter to what we might have expected, Angus lived to tell the tale. It is no good saying that the venom was not poisonous, or that the relevant power was missing, because the scenario makes clear that both were the case. They are, after all, why Angus required treatment, and why he suffered paralysis. Nor is it the case that the relevant power failed to meet with the appropriate conditions for its manifestation. Angus lacks any natural immunity to Tiger Snake venom, and we can safely assume that he had not previously done anything to make his body otherwise unsusceptible to its deadly effects. The power was present, as were the conditions for its manifestation, hence the problem can only lie with the manifestation itself. What can be said about that?

On the one hand, it seems perfectly reasonable to say that the manifestation *did not* come about. This is the assumption Bird relies on when claiming that antidote cases undermine the revised conditional analysis: the power remains and the object receives the appropriate stimulus, but the expected response does not arise (Bird 1998, p. 228). And this seems obvious enough: Angus's death was *prevented* by his taking the anti-venom. But on the other hand there is something altogether too clumsy about saying the manifestation did not come about. There is a perfectly reasonable sense in which the manifestation *did* come about. To say it did not suggests that the poison did nothing at all. But that is mistaken, given that Angus suffered some degree of paralysis in addition to sweating and laboured breathing. Had Angus been immune to the poisoning, we could state unreservedly that the manifestation did not come about, but saying the same of the anti-venom case obscures the contrast between it and the null reaction in the immunity case. What we want to say—one way or another—is that the manifestation both *did* and *did not* come about. But unless we want to embrace dialetheism, we shall have to find a better way of saying it. Finding the best way to say it should help us in our thinking about manifestations more generally.

Before going any further, I want to respond to an early worry concerning this case. If you are anything like me when it comes to thinking about powers, you might be hesitant to treat poisons as the sort of powers that belong in your ontology. In fact, if you are exactly like me, you might flat out deny that there is any such power, claiming that something's being poisonous is a useful ascription of potential object behaviour that is derivative of the genuine powers, and therefore not powerful itself.

[6]Not directly anyway. Certain consequences regarding that debate can be drawn from what I argue.

Though there is a close connection between the two (it is sometimes that of identity), it is important that we distinguish our ascribing various powers to objects from the powers those objects possess (because just as often it is not). For instance, we would not want to *ascribe* the power of fragility to a glass that is forever under the protection of an otherwise unoccupied angel. Under no actual cases will it ever break. That is not to say it fails to possess the relevant power, just that it will never be manifested. We know that different things could occur if the angel was absent, but she never is, so we treat the glass as non-fragile. Conversely, we often say of glasses—truthfully—that they are fragile. But on most accounts of properties, a glass's being fragile is not a matter of its having a property that is its power of fragility. It is true of the glass that it is fragile (to wit, the predicate genuinely applies), but its truth-maker is that the glass has some set of underlying power properties concerning its structure that are responsible for its breaking, if or when it does. And unless those powers are similarly undergirded by other powers, these are the object's genuine powers.

I am not here interested in an analysis of what it is to say that an object will tend to behave in some way or other.[7] I am interested in the sort of manifestations power properties produce. So why concern ourselves with a poisoning case, which is arguably an instance of the former? The short answer is that cases like this provide the best window into how the genuine powers might work. If would-be powers like this are not at all illustrative, then we have little chance making progress in our thinking about the real cases, and condemn ourselves to scepticism about their natures. We are modelling as best we can. That said, when we do so we must be careful not to treat them as always being the same. Would-be powers provide lessons, but we must handle them carefully. As it happens, part of what I think we get out of the poison case is itself informative about the difference between genuine powers and merely true power ascriptions. Thus we can proceed, albeit cautiously.

6.4 Manifestations as Processes

Let us return to the snakebite and the matter of how best to navigate our desire to assert the prima facie contradiction that the manifestation both did and did not come about.

A natural enough proposal is that thinking of manifestations as processes offers us the sort of flexibility we require. According to Handfield, the manifestation of a power "typically—perhaps always... involves a process" (Handfield 2010, p. 106). Ellis maintains much the same, claiming that the display of a power is "a certain natural kind of causal process" (Ellis 2001, p. 124). If the manifestation associated with being a poison is a process whose terminus is death, it is understandable why

[7] Not primarily anyway: something of an account emerges in terms of 'process initiators', but is not developed in detail.

we would want to say both that the manifestation came about and that it did not. Processes—protracted, structured events—paradigmatically involve a series of changes. In this case the final change is that of changing from living to being dead. That change has not come about, hence our negative impulse is explained. However, some portion of the process has come about (changes have taken place, just not the final one), and this is what fuels our desire to claim that we did get the manifestation after all. Furthermore, approaching manifestations in this way provides us with the resources to neatly capture how the anti-venom case differs from immunity cases: the former involves a series of changes brought about by the power in question, whereas the latter null-case involves no such poison-produced changes.

At a first pass this proposal seems to be kosher, and gives us everything we want to get out of the scenario. But I fear that not all is as it first seems. Let us start by clarifying the proposal. If manifestations are processes, then when the right constellation of powers obtains, we get a process. But this is still somewhat vague. There are many ways in which a constellation of powers might *result* in a process (to wit, if a series of events line up the right way), but which fall well short of any single manifestation being a process. For manifestations to be processes it cannot be that a process merely supervenes on whatever else takes place; the manifestation cannot, as Handfield states, merely "involve" a process—the manifestation itself must *be* a process.

An illustration may help. Imagine a simple Turing machine that travels ten spaces forward on the strip of tape, without any movements in any other direction. The machine's ten-space journey is a process, but it can come about in different ways. The first is direct: the only instruction the machine receives is to move ahead ten spaces. During the machine's movement it is not open to further instruction. The second is indirect: the machine reads the tape ten times, following the instruction to move one space each time. A ten-space journey supervenes on the repeated reading and moving, but there is no ten-space instruction. Hence, in the second case the journey is punctuated by points at which the machine is open to new instruction; the journey has causal 'cleavages' at which different instructions could have been inserted. Unlike the first case, in the second the journey could have differed at nine points. To understand a manifestation as a process is for it to be non-supervenient, like the direct version of the ten-space journey.[8]

But now it is no longer clear how this helps us with the anti-venom case. What was initially attractive about treating this case as a process was that as a structured entity its cleavages allowed us to distinguish partially completed processes from those that run to completion or never start. But now this is denied us once more. Manifestations as processes looks to be an all-or-nothing affair: they either come about or they do not. It is their structure that makes them appealing, but this proposal undermines that feature. If we are to exploit the cleavages, we need a means of accessing them. To reiterate: it may well be the case that the exercising of powers 'gives rise to' processes, in the sense that certain processes supervene on multiple

[8]This is discussed in more detail in Williams (2014).

powers producing manifestations over time, but this is not the same as some single power manifesting a process along with its various power partners. That would not allow for the cleavages we seek to exploit, and is open to objection.

Furthermore, though it is possible to interpret 'getting the right constellation of powers required for manifestation' as indicating something about the diachronic distribution of powers, this is a mistaken reading in this instance. Assume, for argument's sake, that being a poison is the sort of power that has to be present throughout a process in order for the process's terminus to be reached.[9] (That is, we deny that being a poison is the sort of power that can start a 'chain reaction' through some initial interaction, but can thereafter be absent.) This means that if the poison meets with the right power partners along its temporal timeline, a certain process will unfold, resulting in death. It would then be tempting to say that the manifestation of the power was the process, partly because the process arose from the power's having met the right partners along the way. I do not deny that this is something we might want to say—it is quite tempting indeed!—but for manifestations to be processes the 'right constellation' must be some set of powers appropriately *synchronically* arranged with the target power, such that the full process springs forth from that meeting. If there are many such meetings, then undoubtedly a process will arise, but not because any one power has as its manifestation a process, but because processes supervene on ever-so-many meetings and manifestations. To wit, statements about powers giving rise to processes would be true, but not because powers have processes as their manifestations.[10]

The objection I have raised to treating manifestations as processes is that this would require that processes come as package deals, and it is clear enough from Angus's survival that they do not. This means that there is no hope of salvaging the manifestations-as-processes proposal by appealing to partial or incomplete processes. But might something be gained by pressing on the multi-track nature of powers? After all, it is a widely held belief that powers can manifest themselves in a number of ways. What stops us treating Angus's antidote case as one in which a different manifestation of the poison comes about? This would be a complete manifestation, but with an alternative terminus.

I am fond of treating powers as capable of manifesting themselves in a variety of ways, but what stops us rescuing the manifestations-as-processes proposal this way is that the alternative manifestation has no way of getting started. If the same power is to produce different manifestations, it has to do so by being exercised in a different constellation of powers. The partners can be different powers, or the same powers differently arranged, but it cannot be that one and the same power produces

[9]If the target power is not present throughout, then the presence or absence of *later* powers cannot be relevant to *its* manifestation.

[10]If those statements concern what processes a given power *would* give rise to, then they would only be true ceteris paribus. You cannot fix which powers a given power will come into contact with over time (not without a laboratory, anyway). This is another reason to reject the proposal. (And is the basic fact that fuels much of the debate over the conditional analysis of power ascriptions.)

non-identical manifestations in identical circumstances.[11] And that is the problem with suggesting that multi-tracking can help us with Angus's poisoning: the circumstances under which the power is exercised *are* those that lead to death. If they were not, he would never have been at risk for death. But intervention was required in order for him to live. This was not an alternative pathway, but the same path cut short. That means it was initiated identically, not in some other way that would have produced some other manifestation.

As a final attempt to resuscitate the manifestation-as-processes proposal, would it help any if we thought of the processes differently? That is, we are treating the manifestation of poison as death. But what if the power in question was not for that process, but something more mysterious, such as a *contribution* to the process? For instance, according to Molnar, the manifestation of a single power is a "contribution": many contributions combine to give rise to the processes we observe (Molnar 2003, p. 195; see also Mumford 2009, p. 104). This suggestion is not without it merits—most notably it would mean that powers would have the same manifestation whenever exercised, regardless of the constellation of partners involved (an attractive feature, for reasons I will not go into)—but it will not help us here. The problem with processes is that they come as complete packages. Treating them as contributions breaks them up, but it breaks them along the wrong dimension. Contributions would be abstract parts of process, but the cleavages we require are breaks along the temporal dimension. Unless contributions are understood as carving processes temporally, then they are of no help. And it is clear that is not how they are understood.

Processes are prima facie attractive for our purposes because they are structured entities. According to Handfield, that structure is *causal* (Handfield 2010, p. 111), and this seems correct. Our cleavages then are causal cleavages, meaning that a process is something of a causal chain: a process is a series of causes and their products extended along the temporal dimension. Given that we are operating within the powers ontology, these causal links in the chain must be the exercisings of powers. If they were not, then we would be dealing with some other sort of causal account, and would be friends of powers in name only. I think this point is often overlooked: a powers ontology is one that replaces or understands causation in terms of the actions of powers *all the way down*. If it turns out that powers are ultimately reliant on something else (some other causal feature, casual black boxes we dare not open, or what have you), then we are not really dealing with powers at all. Hence as friends of powers our processes give way to a structure of powers and their exercising, and so it cannot be that powers have processes as their manifestations.

Perhaps the defender of manifestations-as-processes will object that we have been coming at this from the wrong direction from the get-go. We have been thinking of manifestations as the outcomes towards which powers are oriented: the effects they produce if given the chance. As such, I have interpreted the manifestations-as-

[11]The identities in question are type identities. The same is true of token identities also, though less interesting.

processes proposal as requiring that a successful stimulation of a power will result in a process—generated *en block*, so to speak—with that specific outcome as it terminus.[12] Hence, given that these processes can be mucked with, and further that the mere possibility of mucking requires casual cleavages, we are left thinking that manifestations cannot be processes. But what if manifestations are not simply the outcomes that powers produce, but something prior to them—something intermediary—and *these* are the processes?

What I have in mind here is a middle step, absent in the standard way of conceptualizing powers and their exercising. The standard model is a three-place model. There is the *power*, the *stimulation* of the power, and the *manifestation* that results.[13] It follows on this model that if manifestations are processes, they arise (if and when they do), as wholes, because that is the only option. But is there not something missing here? Where is the part where the power *does* something? Where, one might ask, is the *exercising*?[14] For instance, when the fire heats the kettle, the fire's power to heat—once activated—results in the kettle's getting hot, but the standard model leaves out the *heating*, the very activity by which the stimulated power of the fire brings about the kettle's getting hot.

Anna Marmodoro (2017) provides us with a distinct conception of powers which does justice to this middle step: as with the standard model, we have *powers* and their *stimulation*, but then things get a little different. The powers in the constellation are seen as 'agent' or 'patient': the agential powers are those that transition to an activated state (not a new power, but a new way of being that self-same power), and act with the passive powers, which likewise transition, to bring about a change in the latter. Hence what counts as the manifestation depends on which side of the agent/patient divide one considers. For both the agent and the patient there is an *activation* phase, constituted by a different state of the self-same power. This is a transition *internal* to the powers, marking the shift from potentiality to actuality. Shifting into the active state is the manifestation of a power: its manifestation is the activity it is for (p. 59). But there is also the change that the patient undergoes; and this too is a manifestation. Hence, unlike the standard model, we get manifestations that are not simply causal termini.

So what can thinking about powers in this way do to save the manifestations-as-processes proposal? Not enough, I think, to keep the proposal afloat, but it gives us more to think about.

The change in the patient property is an 'outcome' type of manifestation, and is perfectly in keeping with the standard model, so we can safely put that to one side. Changes are all but universally treated as events, and nothing about these ones makes

[12]Not without good reason: some defenders of the manifestations-as-processes view are committed to exactly this. See, for instance, Ellis (2001).

[13]More complicated versions add extra places for times (and perhaps more), but nothing gets added that undermines the standard conceptual structure as presented.

[14]I think the 'one' here doing the asking is Anna Marmodoro (2017). She would rightly complain that the standard model omits an *activational* feature of the classical (Aristotelian) model, and that it needs putting back. See also Marmodoro (Chap. 5, this volume).

them seem like they would be sufficiently protracted or complicated as to constitute processes onto themselves.[15] If they are short-lived events, then we will deal with them below. That leaves us with the activated powers. According to Marmodoro, the activities of activated powers are distinct events that interact: "the stove *heats* while the pot *is heated*" (Marmodoro 2017, p. 72). This is our middle step, ignored by the standard model, but still not, as far as I can tell, beneficial in rescuing the manifestations-as-processes proposal.

Firstly, there is something odd about a power (or any property) having a different status of the sort we are considering. This is not a question of its being instantiated or not—that sort of change in status is familiar enough—this is an instantiated property having different ways that *it* is, owing to what powers it is partnered with. Though we contrast the potential with the actual, these powers are always actual in that they are there, and so having them change in status when active cannot be quite that. Inactive powers are not unreal powers, which means that change in status must be something like a property of the power properties. I, for one, am not fond of ontologies wherein properties bear properties. Successful or not, this strikes me as a cost of this sort of view.[16]

Secondly, and more importantly, even if we can get past the oddities here, we are still dealing with manifestations that can be mucked with, so we are right back to needing the same causal cleavages we did before. If the fire's power to heat is activated along with the kettle's power of heating up, we can still come along later and add liquid nitrogen to the kettle, thereby preventing the heating. This shows that the activation state of neither power is a prolonged process. Whatever they are, they must be short-lived (and repeated), in order to afford the causal cleavages that provide conceptual space for interruption. They might still be events, or perhaps properties or states, but thinking of the manifestations here as processes looks mistaken. Of course, there is nothing wrong with *saying* that a process-like structure supervenes on the activations and inactivations of the powers involved. If we track the scenario just envisioned, we can follow the placing of the kettle on the fire, the heatings, the adding of liquid nitrogen, and then the coolings—and see that it all flows in a network of interwoven processes, but the causal structure of the world demands that the interweaving that generates the flow is a causal structure of many shorter-lived goings on.[17]

Before we move away from treating manifestations as processes, it is worth pausing to dwell on the conclusion of this section. Despite rejecting the proposal, it remains the case that in a 'loose-speak' sort of way we can describe the powers as having processes as manifestations. But do not now trick yourself into thinking that

[15] For any properties we ought to take at all seriously, there is no halfway house between an object's having them and not having them, so these final changes, when they do occur, ought to be brief. Lombard (1986) is the *locus classicus* for treating changes as events.

[16] Those who favour an abundant view of properties might not find this as costly as it sounds to my sparsist ears.

[17] A Leibniz-ised version of this would escape this worry. But it would do so by jumping out of the kettle and into the fire of causal non-interaction.

manifestations are processes just because it is true that processes are manifest when powers are exercised. Powers are, so to speak, 'process initiators', in the sense that their manifestations determine the constellations from which the next manifestation (s) will be produced, and therefore can be thought of as the first step of a process. *Ceteris paribus*, we can be confident that a certain constellation of powers will favour some ultimate end state many steps down the line.[18] But these processes can be interrupted, blocked, or diverted. It follows that some true power ascriptions—stated *ceteris paribus*—capture the nature of certain objects, and are handy in science and generally useful for navigating the world around us, but do not pick out the powers that do the work. This is not to deny that dried out hunks of clay are brittle, but what makes it true that such a hunk is brittle is not its possessing a power of brittleness, but that it has some constellation of powers that (if allowed to produce their manifestations through sufficiently many iterations without being disturbed) would give rise to a series of constellations and manifestations on which a process like *crumbling* would supervene.[19]

Long story short, treating powers as having short-lived manifestations allows us to say everything we want to about interruption cases, without being forced into error. Hence we should embrace the fact that the exercising of constellations of powers give rise to processes, but not because the manifestation of a power is itself a process or a contribution to a process, but because manifestations can be lined up, and those line-ups are processes. This all breaks down into smaller segments, and that is where we need to be headed.

6.5 Manifestations as Events, States, or Properties

To the extent that my aim was to argue that manifestations must be short-lived, we seem to have come that far. From this point on the task is that of selecting which of the remaining three categories best suits the class of manifestations. Let us start with the events category.

I hazard that when it comes to finding the right category for manifestations, the events category is the default response. It has long been the case that folks have spoken of the displays of powers by using the language of events. We are told that powers manifest themselves in dissolvings, stretchings, burnings, shatterings, breakings, meltings, and the like. This view had its beginnings in an empiricist ontology wherein powers were spooky hidden things and events were a trustworthy visible alternative, but even after that ontology was abandoned the examples stuck. And

[18]Note that we can best make sense of these ceteris-paribus requiring features of processes if we see the processes as *sequences* of exercising powers.

[19]E. J. Lowe (2006, p. 170) argues that the manifestation of brittleness is "essentially a process." If 'brittleness' names a power, then this is strictly false, but it can nevertheless be truly predicated of dried out hunks of clay, in virtue of the powers they do possess. Consequently, if a process is essentially connected with being brittle, then brittleness cannot name a genuine power.

surely with good reason too: many of the terms we apply to objects speak of capacities or liabilities that we observe through their behaviours. But we know better than to be misled by language (everyday loose talk especially), and have already seen that many of the powers we ascribe to objects do not line up with their actual powers. Therefore, though we might correctly describe a portion of salt as being soluble, it does not require that the salt possess any power property so named whose manifestation is the event of dissolving. Nor does it require us to treat manifestations as events, though we might choose to nonetheless.

Events come in all shapes and sizes. Events can be brief—instantaneous even—but they can be much, much longer too. A 90-min soccer match is an event, but so too are the final blast on the referee's whistle and the penalty kick scored late in the second half. Whatever the proper understanding of events is, it must be able to accommodate this potential for great variability in event duration. But as we shall see, it is this same variability that makes the event category an unsuitable for serving as the right category for manifestations.

We have already found the process category unfit for hosting manifestations because processes are structured entities. Processes admit of a causal structure, and this demands they break up into shorter-lived sequences of power-constellation-and-manifestation pairs. Thinking now of longer events (soccer matches, for example), why should these fare any better? Lengthy events are structured entities, composed of ever-so-many changes. These changes betray their causal structure, so we find that longer events are in the same boat as processes: they break down into a series of exercisings of powers, and so cannot be the right category for manifestations.

But of course not all events are lengthy ones; plenty of events are much briefer. In fact on some analyses of events, an event can be very brief indeed. This is not simply a matter of reducing the time involved from that of a soccer match to a sugar cube's dissolving, but getting right down to events that are instantaneous or near enough. For instance, if an event is just the matter of there being some property instantiated at a time, or some change in property, then some events can be incredibly brief indeed.[20] What is wrong with manifestations finding a home here, with events like this?

The short answer is that there is nothing wrong with manifestations being events like this. When events are understood in this way, they can be so brief as to no longer be structured. They are short enough to be the basic casual units that provide the structure of the longer events and processes. This is not to suggest that they are instantaneous—they might be that brief—but nothing dictates that they must be so. They will only be as brief as the exercising of powers happens to be. Whether that is instantaneous is something I am in no position to say, but as the mechanism of change *is* the exercising of powers, it follows that it will be just as brief as any

[20]See Kim (1976) for an analysis of the first sort, and Lombard (1986) for an instance of the second. For reasons we need not enter into here, I treat the 'change' condition as including cases where the exercising of a power results in the instantiation of a type-identical property with that which produced it. In short, my 'changes' are just occasions on which constellations of powers are exercised.

possible change can be, which is brief enough. Nonetheless, I do not think the event category is the right category for manifestations.

The reason I think we should reject the manifestations-as-events proposal is not a pressing one. As I indicated early on, this is an account I would be happy enough with, and treat it as a closer runner up to states of affairs. I only suggest we pass on it because there is an option that ought to make us even happier. Consider the events category as a whole: some events are structured (in the sense we have been discussing), and some are not. But I have argued that no manifestations are ever structured, so we have something of a mismatch here. The significance of this mismatch can be reduced if manifestations constitute a mere subset of the category of events, but we can claim this only if we are willing to admit events into our ontology that are not causally relevant and that supervene on the genuinely causal events, which we might not all be willing to do. Nonetheless, if there is a mismatch, it makes the manifestations-as-events proposal *misleading* to a large extent.

Recall the list of examples that made the proposal initially attractive: dissolvings, stretchings, burnings, shattering, breakings, meltings, and so on. Every one of these examples belongs to the structured half of the events category, the dubious half to which the manifestations do not belong. Consider dissolving: the dissolving of the salt in the hot water is not a one-step event, but rather the product of a series of interactions—a chemical reaction if you will. We can see that it is structured because it is *interruptible*: there are cleavages in the event where the event can be blocked or prevented. Interruptibility is the hallmark of structure, and all the examples bear it.[21] That is why I prefer we find an alternative category for manifestations. The manifestations-as-events proposal is motivated by examples that the manifestations themselves cannot live up to! Moreover, *paradigm* events are structured, so not only is the motivation misguided, the natural way of thinking about manifestations—if we take them to be events—is largely misleading, if not outright mistaken. With the problems starting to mount up, why not put manifestations in a category that does not have the unwanted cases? Why not indeed.

If the events category is not the best home for manifestations, that leaves only properties and states of affairs. Mumford (2009) has described the picture of causation within the powers ontology as the "passing around of powers."[22] Read literally, this would entail that manifestations are properties. When a power is exercised, the result is some property or other. As Mumford indicates, a "fragile glass manifests itself in breaking, for instance, and the pieces now have a power to cut," something the pieces previously lacked (Mumford 2009, p. 98). So are manifestations just properties?

I suggest not. Again, the reasons are not knockdown, but they gently push us in a different direction. The first is that virtually every example we have considered, regardless of the category it seemed best suited for, included not only properties in the manifestation, but some property bearer. It is the *glass* that is shattered, the *salt*

[21]For more on this on the notion of 'interruptibility', see Williams (2014, 2017).

[22]It is not clear that Mumford endorses the literal reading, appearances to the contrary.

that is in solution, the *rubber band* that stretches, and so on. No-one need deny that the exercising of powers gives rise to properties (most likely further powers), but why should we think the manifestation ends there, with just the properties and nothing else? In the example above, it is not the case that the manifestation results in the power to cut being instantiated *simpliciter*, but that the pieces of glass have this property. Perhaps this objection can be answered by saying that not only are powers manifested in properties, but part of the nature of exercising and manifesting is that it determines the locations for these properties too. But this only helps if there is nothing more to property bearers than the properties themselves. Do we really want our account of manifestations to force us to be bundle theorists?

Let us assume that we are not so inclined as to adopt the bundle theory merely to save the manifestations-as-properties proposal. The problem remains that we seem to require property bearers for the properties that are the manifestations. But this has yet another unsavoury consequence: treating manifestations as properties cuts objects out of the causal nexus altogether.

Whatever you take the real objects of this world to be, it is hard to deny that their comings and goings are a causal matter. If something comes into being or ceases to be, it is surely the result of powers being exercised in this or that way. I am also of the opinion that their continuing to be is a causal matter. But unless one is willing to endorse the incredibly unpopular thesis that 'existence' names a property (I am not!), then either there has to be an independent and non-powers based account of the causal story here, in which case our powers theory is not the whole story, or objects are a part of the manifestations powers produce. I opt for the latter.

Hence, treating manifestations as properties forces us into something of a dilemma regarding the bearers of those properties. On the one hand we can accept that manifestations are properties and nothing more, and avoid the problem of property bearers by taking substances themselves to be nothing more than properties. On the other hand we can avoid treating substances as bundles of properties only if we are willing to concede that when those substances come into being, persist, or go out of being, they somehow manage to do this *outside* of the powers-based causal network. That is, we either endorse the implausible thesis that none of this is a causal matter, or we buy into some non-powers based account of causation for substances. I cannot say I have any clue what that would look like, but even if I did, I have a hard time imagining how the distinct causal mechanisms would mesh, or why we want to bother with powers any more.[23]

Instead, I suggest we let the powers do all the work. That is what they were built for, and a significant part of why theoretical entities of this sort were hypothesised in the first place. Letting them do their jobs means making substances part of manifestations. We should treat manifestations as states of affairs.

A state of affairs is a matter of some substance's having some property or other or standing in a relation to some other substance. They can be instantaneous, but nothing stops them from being longer than this. If we treat manifestations this

[23]See Groff (Chap. 3, this volume) for more on this issue.

way, we get to keep what is attractive or intuitive in thinking of them as events or just properties, but avoid any of the minor worries we have with those conceptions. We also get a tidy way of capturing out intuitions about causal processes, allowing us to explain them entirely in terms of powers and their being exercised as parts of constellations, without running afoul of interruptibility.

Consider the poison case again, and our desire to say that the manifestation both did and did not come about. We can now see it for what it really is, and why it was we were so keen to assert something impossible about the process. The ingestion of the poison is a matter of the power of the poison interacting with a constellation of other powers within the body. That resulted in another state of the body's constituents, which in turn resulted in yet another. Many iterations of constellations of powers giving rise to further objects with further powers are followed by yet more constellations and yet more exercised powers. If left alone—to wit, if nothing else is added to the constellation—then the series of constellations will eventually give rise to a state of affairs that is Angus's death. But the series of constellation-manifestation pairs is mucked with, when the antidote is added to the mix. This new reagent adds powers to the constellation with a different sort of manifestation—one that does not lead to death.

As I have stressed, we are free to see the processes as a whole as supervening on the exercise of powers, and I recommend that we do. But note that no process ever really starts or ends, it is just one little constellation, then the next. Some sequences of manifestations we track, others we do not. A process we see as derailed is really just an expected sequence that is continued I an unexpected or undesirable direction. No wonder some friends of powers speak as if all powers-based causation is a network of interacting processes! And now we know how that can be so, by treating manifestations as they ought to be understood.

Of course, this new appreciation of what manifestations are and how processes emerge has something of a minor cost. Recall bullet (2) from above, that we tend to name powers by way of the manifestations they produce. Though something like this could still hold, it would not line up with our standard ascriptions. Poison (properly understood) can result in death if ingested, but not because that is its manifestation, but because it sets up a sequence that will include death, if not mucked with. Hence, most of the powers named for their effects are not named for their manifestations at all, but for states that come at the end of a sequence. That is why it is convenient to think of powers as processes initiators (if not strictly correct), because it continues to be useful to track states that can come about when powers are manifested, even if those manifestations are not produced directly by the power in question. Hence, though it might be a minor cost of this view (and any that treats manifestations as brief or instant) that standard ascriptions are rarely identical with genuine power properties, they nevertheless remain advantageous for science and world navigation, because they connect with the real powers in the right sort of way.

But there are also benefits of manifestations being states of affairs, beyond a tidy treatment of processes. For instance, there are a number of regresses that have been raised against treating powers as 'pure' powers: that is, as properties whose nature is exhausted by their ability to produce their manifestations. It has been claimed many times over that such a conception of powers is not viable on the grounds that if all properties are powers, and there is nothing more to them than their powerful nature, then there is nothing more to the power *P* than its power to produce *Q*, and *Q* is nothing more than the power to produce *R*, and *R* is nothing more than the power to produce *S*... and so on.[24] At no point do we get a manifestation that is not itself a pointing at a yet further manifestation. Thus objectors claim that the properties have no identities because there is nothing to fix them. Solutions to the problem have been offered, and it is safe to say that the regress is no longer considered nearly as worrisome as it once was.[25] But note that the regress never gets going unless one endorses the manifestations-as-properties proposal. If a manifestation involves something more, that is, if a manifestation is a matter of *some thing's being some way*, then *P* is the power for *x*'s being *Q*, and its nature is fixed, even if there is nothing more to *P* than its powerfulness.[26]

6.6 Conclusion

In his oft-cited response to Martin's rejection of the simple conditional analysis, Lewis writes:

> Sometimes it takes some time for a disposition to do its work. When stimulus *s* arrives and the disposition is present, some process begins. (It might be a process of accumulation: of charge, of neurotransmitter, of tiny cracks, of vexation, etc.) When the process reaches completion, then that is, or that causes, response *r*. But if the disposition went away part-way through, the process would be aborted. (Lewis 1999 p. 137).

I think Lewis here is mistaken in all sorts of ways. If the aim here is that of rightly capturing our everyday power 'talk', then perhaps he can be forgiven. But if this about the metaphysics of powers, then he is well off the mark.

He is right that *r* has not really come about. And he is right that a process begins. But he is mistaken in thinking that processes get aborted if the relevant power goes missing: some powers might work this way, just as many do not. And more importantly, if I am to be believed regarding how best to understand manifestations, then he is mistaken in his claim that powers take some time to do their work.

Powers do *never* need time to do their work. They do it as quickly as is possible. That is not to say they do it instantly: it takes just as long as is needed, but it is so brief that there is no space to interrupt them, not even in principle. And then they might do some more work right afterwards, perhaps as part of a constellation very similar to the first. And then they might do more work—again and again—in a series

[24] See Swinburne (1980) for what is perhaps the first argument of this sort. I am not here claiming that powers are exhausted by their powerful nature, but many friends of powers do.

[25] See Bird (2007) for one such response.

[26] Heil (2003, ch. 10) argues against the pure powers thesis, but offers a subplot similar to the argument I have raised.

of constellations, all of which might differ somewhat. And we might even come to see this series as a process in which the power is a key constituent.[27] But Lewis would still be mistaken in thinking that powers take time to do their thing. Powers act quickly, and give rise to states of affairs. Simply put, manifestations are brutish and short—and when poisons are involved, can be quite nasty too.[28]

Acknowledgements Thanks to Anna Marmodoro, Anne Sophie Meincke, and audience members in Innsbruck, for many helpful comments.

References

Armstrong, David. 1997. *A World of States of Affairs*. Cambridge: Cambridge University Press.
Bird, Alexander. 1998. Dispositions and Antidotes. *The Philosophical Quarterly* 48: 227–234.
———. 2007. *Nature's Metaphysics*. New York: Oxford University Press.
Carnap, Rudolph. 1928. *The Logical Structure of the World*. Berkeley: University of California Press.
———. 1936–1937. Testability and Meaning. *Philosophy of Science, 3*: 420–471 and 4: 1–40.
Choi, Sungho. 2018. Dispositions, *The Stanford Encyclopaedia of Philosophy* (Fall 2018 Edition), ed. Edward N. Zalta. https://plato.stanford.edu/archives/fall2018/entries/dispositions/. Last accessed on 10 May 2019.
Ellis, Brian. 2001. *Scientific Essentialism*. New York: Cambridge University Press.
Groff, Ruth. This volume. What Does the Doing? On Powers, Things, and Powerful Things. In *Dispositionalism. Perspectives from Metaphysics and the Philosophy of Science (Synthese Library* 417*)*, ed. A. S. Meincke, 27–40. Cham: Springer.
Handfield, Toby. 2010. Dispositions, Manifestations, and Causal Structure. In *The Metaphysics of Powers – Their Grounding and their Manifestations*, ed. A. Marmodoro, 106–132. New York: Routledge.
Heil, John. 2003. *From an Ontological Point of View*. New York: Oxford University Press.
Kim, Jaegwon. 1973. Causation, Nomic Subsumption, and the Notion of Event. Reprinted in his *Supervenience and Mind*, 1993, New York: Cambridge University Press, 3–21.
———. 1976. Events as Property Exemplifications. In *Action Theory (Synthese Library* 97*)*, ed. M. Brand and D. Walton, 159–177. Dordrecht: Springer. Reprinted in his *Supervenience and Mind*, 1993, New York: Cambridge University Press, 33–52.
Lewis, David. 1997. Finkish Dispositions. In his 1999 *Papers in Metaphysics and Epistemology*. New York: Cambridge University Press, 133–151.
———. 1999. *Papers in Metaphysics and Epistemology*. Cambridge: Cambridge University Press.
Lombard, Lawrence. 1986. *Events – A Metaphysical Study*. London: Routledge.
Lowe, E. Jonathan. 2006. *The Four-Category Ontology*. New York: Oxford University Press.
———. 2010. On the Individuation of Powers. In *The Metaphysics of Powers – Their Grounding and their Manifestations*, ed. A. Marmodoro, 8–26. New York: Routledge.
Marmodoro, Anna. 2017. Aristotelian Powers at Work: Reciprocity Without Symmetry in Causation. In *Causal Powers*, ed. J. Jacobs, 57–76. New York: Oxford University Press.
———. This volume. Powers, Activity and Interaction. In *Dispositionalism. Perspectives from Metaphysics and the Philosophy of Science (Synthese Library* 417*)*, ed. A. S. Meincke, 55–66. Cham: Springer.

[27]See Meincke (Chap. 7, this volume) for an alternative take on powers and processes.
[28]Many of the ideas in this chapter get further development in Williams (2019).

Martin, Charles B. 2008. *The Mind in Nature*. New York: Oxford University Press.
Meincke, Anne Sophie. This volume. Powers, Persistence and Process. In *Dispositionalism. Perspectives from Metaphysics and the Philosophy of Science (Synthese Library* 417*)*, ed. A. S. Meincke, 89–113. Cham: Springer.
Molnar, George. 2003. *Powers*. New York: Oxford University Press.
Mumford, Stephen. 2009. Passing Powers Around. *The Monist* 92: 94–111.
Mumford, Stephen, and Rani L. Anjum. 2011. *Getting Causes from Powers*. New York: Oxford University Press.
Ryle, Gilbert. 1949. *The Concept of Mind*. London: Cambridge University Press.
Swinburne, Richard. 1980. Properties, Causation and Projectibility: A Reply to Shoemaker. In *Applications of Inductive Logic*, ed. L.J. Cohen and M. Hesse, 313–320. Oxford: Oxford University Press.
Williams, Neil E. 2011. Putting Powers Back on Multi-Track. *Philosophia* 39: 581–595.
———. 2014. Powers: Necessity and Neighborhoods. *American Philosophical Quarterly* 51: 357–371.
———. 2017. Powerful Perdurance: Linking Parts with Powers. In *Causal Powers*, ed. Jonathan Jacobs, 139–164. New York: Oxford University Press.
———. 2019. *The Powers Metaphysic*. Oxford: Oxford University Press.

Chapter 7
Powers, Persistence and Process

Anne Sophie Meincke

Abstract Stephen Mumford has argued that dispositionalists ought to be endurantists because perdurantism, by breaking down persisting objects in sequences of static discrete existents, is at odds with a powers metaphysics. This has been contested by Neil Williams who offers his own version of 'powerful' perdurance where powers function as links between the temporal parts of persisting objects. Weighing up the arguments given by both sides, I show that the profile of 'powerful' persistence crucially depends on how one conceptualises the processes involved in the manifestation of powers. As this turns out not to be determined *per se* by subscribing to some view labelled 'powers view', further discussion is needed as to what processes are and to what kind of process theory a powers metaphysics should commit itself in order to be convincing. I defend the claim that dispositionalism is best combined with a version of process ontology that is indeed incompatible with a perdurantist analysis of persistence. However, I argue that this does not imply that dispositionalists ought to be endurantists.

Keywords Powers · Dispositionalism · Persistence · Perdurance · Endurance · Process

7.1 Introduction

Reality as we experience it in daily life is characterised by a copresence of identity and change. Nothing lasts forever; however, things like tables, trees, thunderstorms, marriages or people do last at least for a certain time. And while they last, as everybody knows, they undergo all sorts of changes. Tables may change their colour, trees grow and shed their leaves, thunderstorms proceed from producing dazzling lightnings on the horizon to uprooting trees and blowing off roofs,

A. S. Meincke (✉)
Department of Philosophy, University of Vienna, Vienna, Austria

Centre for the Study of Life Sciences (Egenis), University of Exeter, Exeter, UK
e-mail: anne.sophie.meincke@univie.ac.at; a.s.meincke@exeter.ac.uk

A. S. Meincke (ed.), *Dispositionalism*, Synthese Library 417,
https://doi.org/10.1007/978-3-030-28722-1_7

marriages start out as passionate love affairs before all too often turning into nasty battlefields, and the people who get caught up in this do not only occasionally change their partners, but also their sizes, moods, looks, jobs, houses and so on.

The debate in metaphysics on persistence is to be understood as the struggle to make sense of this copresence of identity and change. David Lewis (1986) famously formulated the challenge like this: how can a thing change its intrinsic properties over time and still be the same, numerically identical thing, given that numerical identity, according to Leibniz's Law, entails qualitative identity, i.e., having exactly the same intrinsic properties? Lewis's answer to this so-called problem of temporary intrinsics is perdurantism: the view that a thing persists "by having different temporal parts, or stages, at different times, though no one part of it is wholly present at more than one time" (Lewis 1986, p. 202). Perdurantism is opposed to endurantism according to which, as Lewis puts it, a thing "persists by being wholly present at more than one time" (ibid.). Following this canonical distinction, in the last three decades ever more sophisticated versions of both perdurantist and endurantist accounts of persistence have been produced.[1] However, the debate is far from settled. It is not clear which of the two approaches is the superior one, if any of these is convincing at all.

In this chapter I investigate how a realist understanding of dispositional properties bears upon the controversy about persistence. What sort of account should dispositionalists endorse? Should they be endurantists or perdurantists? Or neither of these? Stephen Mumford (2009, p. 226) has argued that while the powers ontology "doesn't quite" "deliver for us a verdict on the endurance/perdurance debate", "a powers ontologist nevertheless has some reason to favour endurantism as sitting most comfortably with their ontology". Mumford dismisses perdurantism for resting upon a Humean ontology that assumes static and discrete existents as key entities. However, Neil Williams (2017, 2019) has challenged Mumford's argument by proposing a version of a perdurantist account where powers function as links between those static existents that are supposed to compose persisting objects.[2] This shows that one cannot simply rely on the traditional anti-Humean stance of dispositionalism when reflecting on the latter's implications for other questions in metaphysics. Instead, it becomes open to debate what the ontological commitments of dispositionalism are or should be.

In what follows I present (Sects. 7.2 and 7.3) and weigh up (Sect. 7.4) the arguments given by both sides against and in favour of 'powerful perdurance'. It turns out that the disagreement between Williams and Mumford about how to account for persistence from a dispositionalist point of view has indeed its ultimate roots in a territory reaching beyond dispositionalism itself: it reflects the antagonism

[1] I regard stage theory (exdurantism) as a version of perdurantism and adverbialism as a version of endurantism.

[2] Meanwhile also Dumsday (2019) has offered arguments in favour of reconciling perdurantism with dispositionalism, published after the completion of the present book chapter. Dumsday does not in detail discuss Williams's proposal.

between a view of reality as fundamentally being made up of static things ('thing ontology') on the one hand and a view of reality as being fundamentally processual and changing ('process ontology') on the other. This antagonism is manifest most obviously in Mumford's and Williams's disagreement on how to conceptualise the causal processes involved in the manifestation of powers. Looking at the respective commitments concerning causation and dynamism (Sects. 7.4.1 and 7.4.2), I defend the claim that the more neo-Humean orientation of powerful perdurance does not per se disqualify it from representing a legitimate version of dispositionalism. However, I also provide some reasons for why still a more preferable version of dispositionalism would be one that does not rest upon a thing ontological view of reality but rather takes reality to be seriously dynamic as espoused by process ontology. This includes raising some doubts as to how good a match dispositionalism and endurantism would be (Sect. 7.5). I conclude by indicating the implications of the results of this investigation for future research on both powers and persistence (Sect. 7.6).

7.2 Mumford's Argument Against (Powerful) Perdurance

Mumford's argument has the following structure:

P1 (*Dynamism Premise*): Perdurantism's commitment to static and discrete temporal parts which supposedly compose persisting entities contradicts the powers ontology's commitment to "continuous, dynamic, natural processes involving active particulars" (Mumford 2009, p. 228).

P2 (*Causation Premise*): The perdurantist's adoption of external and contingent relations by which the discrete temporal parts are bound together falls short of a "satisfactory account of causation by the lights of the powers theory" (Mumford 2009, p. 226).

C: "If one were to get an account of persistence from a theory of powers, therefore, it would be an endurance theory" (ibid.).

Two things are noteworthy.

First, the argument is merely negative: it does not tell us positively why we, as dispositionalists, should endorse endurantism; instead, it only tells us why we should *not* endorse perdurantism. Hence, even if it proved to be true that dispositionalism does not fit well with perdurantism, it may still not be true that it does fit well with endurantism. Only under the assumption that the alternative between perdurantism and endurantism is exhaustive could the conclusion **C** appear to be conclusive; but as we shall see later, this assumption can be contested.[3] Mumford's case for endurantism is, at best, incomplete. However, this is not to say that Mumford's argument may not succeed in supporting, rather than **C**, the weaker thesis

[3]Even if granting it for the sake of argument, dispositionalism could still turn out to be incompatible with endurantism. Maybe dispositionalism does not deliver any coherent and convincing account of persistence at all?

C*: If one were to get an account of persistence from a theory of powers, it would not be a perdurance theory; i.e., dispositionalists ought not to be perdurantists.

Second, the argument appeals to ontology. Perdurantism and dispositionalism are claimed to incompatible with one another as they come with incompatible ontological commitments.

According to **P1** (which one may call the 'dynamism premise'), there is a clash between dispositionalism and perdurantism insofar as "perdurance theory has an ontology in which the key entities are unchanging or static and discrete" (Mumford 2009, p. 226). The perdurantist's key entities Mumford has in mind are the perduring thing's temporal parts which, as he correctly reminds us, "have to be unchanging [...] in order to solve the problem for which they were designed", namely Lewis's problem of temporary intrinsics: "so as not to ascribe incompatible properties to the same thing, the theory ascribes them to different temporal parts of the thing" (ibid.). Moreover, following Hawley (2001) Mumford adds that, given the constant possibility of change, temporal parts have to be "as fine-grained as time itself" (Mumford 2009, p. 227). Each thing thus consists of "a large number of changeless and fleeting temporal parts" (ibid.). This, he argues, is in sharp contrast with the dispositionalist picture, in which "the world's particulars [...] have within themselves a principle of change" rather than "being a passive recipient of change": "if a particular has a disposition then it intrinsically tends towards some kinds of process rather than others" (Mumford 2009, p. 228). These processes, Mumford insists, are continuously changing in the sense that they contain "no proper portion that is not undergoing change" (ibid.). By breaking such processes down in "static, instantaneous parts" (ibid.), perdurantism reduces dynamic properties, such as heating, rotting or growing, to sequences of static properties, which are "taken as primarily real" (Mumford 2009, p. 230).

This takes us to **P2**, the claim that perdurantism entails a view of causation unacceptable to the dispositionalist ('causation premise'). Following Ellis (2001), Mumford regards as an integral part of the dispositionalist view the assumption "that there are natural kinds of processes, which have their properties or stages essentially" and that "there are some natural kinds of thing [...] that would not be the things they are unless they partook" in certain kinds of process (Mumford 2009, p. 228). This sort of essentialism entails that the sequence of stages of any such natural process is not contingent. Instead, each stage is intrinsically and necessarily disposed towards the next one by virtue of its being part of a certain kind of natural process (Mumford 2009, p. 231). Perdurantism, in contrast, by giving ontological priority to unchanging discrete entities, has to resort to external, non-supervenient relations to explain how some such unchanging discrete entities form a sequence, i.e., "are temporal parts of the same thing" (Mumford 2009, p. 229). Perdurantists may try to sell these external relations as "everyday causal relations" (Mumford 2009, p. 226), but this clearly will not satisfy the dispositionalist who precisely objects to this view of causation as something owing its existence to a "further fact, over and above the intrinsic facts about the relata" (Mumford 2009, p. 229).

For the reasons expressed in **P1** and **P2**, Mumford also dismisses a hypothetical "mixed view" "in which the requisite non-supervenient relations are provided by the real causal powers of dispositionalism" so that "each static temporal part or stage might have an added power towards the next part or stage" (Mumford 2009, p. 231). First, as being attached to static single stages rather than being embedded in a continuous process, powers would lack, according to Mumford, "the constraints brought by the dispositional essentialist commitment" (ibid.). As a result, it would still "be entirely contingent what static properties went with what powers" (ibid.). Second, since each stage is supposed to be static, there seems to be no ontological basis for the next stage in the series to be causally produced by the previous one unless there were "'static powers' which could be instantiated at an unchanging instant" (Mumford 2009, p. 232). However, Mumford argues, it is hard to conceive of such powers as truly active (which they should be in order to cause something, i.e., to be sources of change), without there being any "grounding in [a] particular's movements or changes" (ibid.). Mumford concludes that perdurantism, as "a variety of Humeanism" (Mumford 2009, p. 233), has to be rejected by the powers theorist even if it were to decorate itself with powers in some way or another.

7.3 Williams's Argument for Powerful Perdurance

As it happens, Neil Williams has put forward a version of perdurantism exactly along the lines of the mixed view envisaged and dismissed by Mumford. The core idea of his theory of powerful perdurance is that "powers serve as the glue that links temporal parts" (Williams 2017, p. 140): "each temporal part of a perduring entity is the manifestation of a specific power of its immediate predecessor: those powers are powers *for* the existence of further temporal parts" (Williams 2017, p. 145; emphasis in the original). Williams claims that objections to this view can be refuted by thinking harder about "how we conceptualize powers" (Williams 2017, p. 140). This means to question "a general conception of powers which makes them look ill-fitted" for the role assigned to them by the theory of powerful perdurance (ibid.).

First, we have to understand that "powers-based perdurance is committed to powers that are not for changes", which implies that "not all causation results in a change, and not all powers need to have different powers as their manifestations" (Williams 2017, p. 147). Instead, persistence, as a causation "for maintaining the status quo" (Williams 2017, p. 146), involves powers that, in some particular sense, "have themselves as manifestations" (Williams 2017, p. 149). The idea is that "properties of one temporal part [are] the cause of that same property's being instantiated by some later temporal part of that same object" just as when we say, "that some object has the shape it does now *because* it had that shape immediately prior to now" (ibid.; emphasis in the original). However, as this formulation already indicates, Williams thinks that one actually should better conceptualise manifestations of powers not as powers, i.e., properties, but as states of affairs so that a temporal part's having a particular property manifests itself in another temporal part's having the same property (Williams 2017, p. 149 f.; for a detailed defence see Williams, Chap. 6, this volume).

Second, Williams asks us to acknowledge that not all powers manifest themselves triggered by the presence of a suitable manifestation partner and, i.e., that not all causation is a case of the mutual manifestation of reciprocal powers. Persistence, according to Williams, is a case of immanent causation (in the sense introduced by Johnson 1964), and as such involves so-called unilateral powers that do not need to be triggered in order to manifest themselves. Appealing to the familiar case of spontaneous radioactive decay, Williams calls for a revision of the common concept of a power in "that the notion of 'stimulus' is not, as had been thought, part of the basic understanding of what it is to be a power, but is rather confined to those powers that are exercised mutually (Williams 2017, p. 154). When it comes to making sense of "powers for the very persistence of particulars" (Williams 2017, p. 153), there is no point in investigating "what conditions must arise for [these powers] to be manifested" (Williams 2017, p. 155) since "persistence does not look to be interactive" (Williams 2017, p. 140). Instead, we should "think of a causal ability" (Williams 2017, p. 155; for a detailed defence of unilateral powers see also Williams 2019, ch. 7.4).

Finally, the "incredibly short, perhaps even instantaneous" duration of temporal parts (Williams 2017, p. 155) entails the same sort of short-livedness for the manifestations of powers (on the latter, see also Williams, Chap. 6, this volume). However, this does not mean that powerful perdurance could not accommodate processes, or that there was some fundamental incompatibility between the dynamicity and continuity of manifestation of powers on the one hand and "the perdurantist's unchanging static temporal parts" on the other (Williams 2017, p. 156). The impression that this was the case – as articulated in the first premise of Mumford's argument against powerful perdurance, **P1** – stems from the presupposition of a particular concept of process which, according to Williams, is neither without alternative nor generally accepted amongst dispositionalists nor in fact the most suitable one for a powers ontology. He argues that once we drop this mistaken concept of process, the alleged incompatibility disappears.

The concept of process Williams has in mind here is what he calls the *adamantine model*, which sees processes "as continuous and impenetrable", i.e., as not dissectible into static, unchanging parts (Williams 2017, p. 157). He concedes: "if processes are adamantine, then they are not amenable to the perdurantist's framework" (ibid.). However, there is also another way of conceptualising protracted processes: the *domino model*, according to which "processes are drawn-out events encompassing a great deal of change" but yet "dissectible into ever-so-many short-lived events, none of which itself includes change" (ibid.). Williams argues that the domino model "fits best how we think about powers and their manifestations" (ibid.). It is generally agreed that powers initiate causal chains that are interruptible as evinced by the possibility of antidotes. Talk of 'causal chains' would be inappropriate if the processes at issue were "seamless", and so would be the assumption of interruptibility for "to be able to stop a process part way through requires that there are natural breaks in that process" (Williams 2017, p. 159).

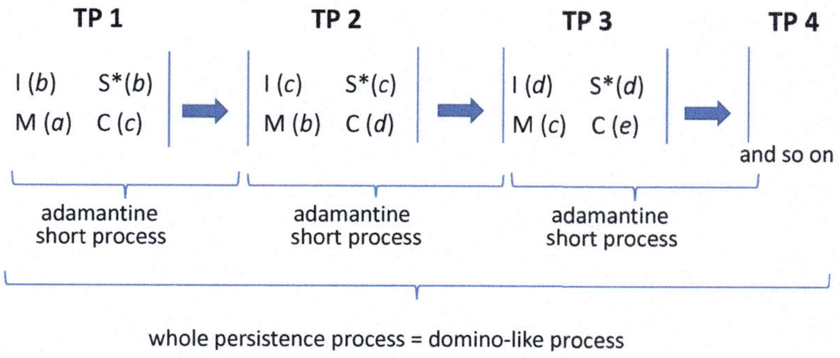

Fig. 7.1 Powerful perdurance

The domino model, if combined with powers, requires us to deny that powers manifest in "*protracted* processes" (Williams 2017, p. 159; emphasis in the original). Because causal chains unfold stepwise, and because "each of these causal steps is due to the actions and interactions of causal powers" (ibid.), it is

> mistaken to claim, in any strict sense, that this causal chain – *in its entirety* – is the product of a single power being stimulated. A more apt description is that the first power stimulated *initiates* a causal chain; to wit, its manifestation is a state of affairs that contains a power for some further state of affairs, that contains a power for some further state of affairs, that. . .and so on until we have a state of affairs that is the 'typical' manifestation we tend to speak of. (Ibid., emphases in the original)

While the "natural cleavages" (ibid.) between the single "power/stimulus/manifestation links" (Williams 2017, p. 160) allow for being interfered with, the individual links "are not interruptible because they do not have natural breaks" (ibid.). They are "adamantine" "short processes" (ibid.).

This basic picture of the "powers metaphysic" envisaged by Williams (see the title of Williams 2019) needs to be adjusted a little for the special case of persistence, given that powers for persistence, according to Williams, result in type-identical manifestations and do not need to be stimulated in order to manifest themselves. This is depicted in Fig. 7.1.

'TP' stands for 'temporal part', 'I' for 'instantiation', 'S*' for 'stimulation analogue', 'M' for 'manifestation' and 'C' for 'relation of containing'. '*a*', '*b*', '*c*', '*d*', '*e*' denote powers for persistence and '→' the causal action of one temporal part bringing about another temporal part. Temporal parts, it thus emerges, "are themselves nothing more than (substances at locations instantiating) constellations of powers" (Williams, personal conversation, July 2013).[4] Looking at TP 1 in the above figure, this means more precisely the following: as the manifestation of some power *a*, TP 1 instantiates the power *b*, and that means it exercises *b*, which would amount to *b*'s being stimulated if *b* was a reciprocal power (powers for persistence are

[4]Williams aims to substitute in general the concept of constellations of powers for the concept of a stimulus; see Williams 2014, 2019, ch. 6.2.

unilateral); and this complex state of affairs contains in itself the power *c* insofar as the power *b* is intrinsically directed towards *c*, i.e., causes it to exist.

Williams insists that the coming-about of a short adamantine process does not take any longer than the duration of any temporal part of the persisting object; there is no need to resort to ever shorter temporal parts (Williams 2017, p. 160). Furthermore, the objection that powers cannot be accommodated in single temporal parts because temporal parts are instantaneous while the exercising of powers is not can be rejected for two reasons: (i) powers might be instantaneous and (ii) temporal parts do not have to be instantaneous.[5] Williams leans towards (i), but points out that for the powers theorist it is natural to think that "when it is powers that are responsible for change, *they* will dictate the duration of temporal parts" (Williams 2017, p. 162; emphasis in the original). Hence, it could still turn out "that temporal parts are instantaneous after all, just as long as that is how fast powers are manifested. But if the powers take a little more time, then the temporal parts will be a little bigger too" (ibid.). There thus is no "mismatch between the duration of temporal parts and the time it takes a power to be manifested" (Williams 2017, p. 161).

Williams concludes that if we revise our concept of a causal power in the three respects presented, then "it can no longer be claimed that powers are not configured in the right sort of way to serve as the glue that links temporal parts into persisting objects" (Williams 2017, p. 164). To wit, if we assume a special class of powers that (i) do not produce change but preserve the status quo, (ii) are unilateral and (iii) result in "incredibly brief" manifestations, then "powers and their exercising are [...] primed to serve as the immanent causal mechanism" (ibid.). Any fears of the succession of temporal parts being a matter of mere contingency are therefore unfounded. It is simply not true, Williams insists, that a perdurance theory of persistence, when combined with powers, is committed to external and contingent relations as binding together temporal parts – which was the second premise of Mumford's argument against powerful perdurance, **P2**. Instead, by deploying powers, i.e., "intrinsic properties with dispositional essences" (Williams 2017, p. 144), the perdurantist can provide exactly those intrinsic and necessary links desired by the anti-Humean powers theorist (Williams 2017, p. 163 f.). The "claim that powers and perdurance are an unhealthy mix" (Williams 2017, p. 163) is mistaken.

7.4 Weighing Up the Arguments: Causation and Dynamism

Has Williams defeated Mumford's argument? Can dispositionalists be perdurantists or ought they even to be such? Williams is cautious enough not to make any recommendations of the latter sort. To defeat both the strong thesis **C** that

[5]Williams (2017, p. 161) rejects Hawley's (2001) abovementioned argument for temporal parts being as fine-grained as time itself on the basis of the assumption that existing at a time is an extrinsic rather than intrinsic property, i.e., a relation in which a temporal part stands to a time.

dispositionalists ought to be endurantists and the weaker thesis $\mathbf{C}^*$ that dispositionalists ought not to be perdurantists it is sufficient to show that dispositionalism can plausibly be combined with perdurantism. Williams believes that dispositionalism is neutral as to the alternative endurantism versus perdurantism (Williams 2017, p. 139), and this is to say that persistence, as a topic in metaphysics, is neutral as to the alternative Humeanism versus anti-Humeanism. It is not that by being a perdurantist one is bound to be a Humean; one can rather adopt a powers view and thereby join the anti-Humean camp (Williams 2017, p. 164).

If this turned out to be true, Mumford's criticism would appear to miss the point by wrongly treating the theory of powerful perdurance as an enemy, instead of welcoming it as an ally in a joint opposition against neo-Humeanism. As will become clear from what follows, Mumford cannot fairly be accused of this. There are reasons to doubt that Williams' theory of powerful perdurance is as anti-Humean as it alleges to be. Mumford's criticism can help us identify these particular elements of Williams's view. However, this will not by itself prove that Williams has failed to defeat Mumford's argument. It remains an open question whether the theory of powerful perdurance rightfully claims to be a 'powers metaphysic'. Williams says so; Mumford disagrees. To settle the case, we must elucidate the respective motivations leading to the diverging ontological commitments on both sides.

7.4.1 Causation

Why do we need causation for persistence? Williams gives a clear answer: "none of the benefits one might gain from cutting up objects into temporal parts can be enjoyed without also providing a means of sticking them back together" (Williams 2017, p. 142). This required means – "the glue that holds object stages together" – must be "causal, as only a causal account is strong enough to give us the right persisting entities" (Williams 2017, p. 143). Powers ensure a non-contingent composition of persisting entities as domino-like processes.

Williams's causal analysis of persistence entails a rejection of a centrepiece of Mumford and Anjum's dispositionalist account of causation: the simultaneous presence of cause and effect. According to Mumford and Anjum, rather than as temporally preceding its effect, we ought to think of a cause as occurring simultaneously with its effect in the sense of "merging into and becoming the effect through a natural process" (Mumford and Anjum 2011, p. 107). The paradigmatic case they appeal to is the 'sweet solution' produced in a temporally extended process from a sugar cube immersed in water. They write:

> It may take time for the sugar to be completely dissolved, but this doesn't mean that the cause and effect are temporally separated. Causation occurs as soon as the sugar and water are united and continues, unless the process is interrupted, until total dissolution has been achieved and then ends at that point. (Mumford and Anjum 2011, p. 107)

Williams dismisses this picture on the grounds that it presupposes a coarse-grained or macroscopic perspective that he thinks is scientifically and metaphysically inadequate. The alleged simultaneity of cause and event, if we are to make any sense of it at all, has to be interpreted as a "wide-scope simultaneity": "that is, the simultaneity will be for *parts* of the stimulation with *parts* of the manifestation, when we consider the entire process over the full extent of its duration, and the manifestation the terminus of that process" (Williams 2014, p. 364; emphases in the original). However, this is not how things really are. Williams explains:

> If we zoom in on the process, the simultaneity is lost: the dissolution of the outermost molecules follows very shortly after their immersion. At no time are the very same molecules both dissolved *and* not dissolved. It is a series of brief (non-simultaneous) steps like this – molecules now in contact with the solvent and then quickly dissolved – that give us the overall process that we treat as (wide-scope) simultaneous. (Williams 2014, p. 365; emphasis in the original)

In other words, the concept of causation as involving a simultaneity of cause and effect can appear to be plausible only under the assumption that there are macroscopic objects with macroscopic powers – and this assumption, according to Williams, is illusory. In his view, there is no simultaneity of cause and effect ultimately because there are no macroscopic powers and no macroscopic objects either in any strict and robust sense.[6] There is no power of sugar cubes to produce a sweet solution when immersed in water, and no power of water to produce a sweet solution when put in touch with a sugar cube because there are, strictly speaking, no such things as sugar cubes and water. What there really is, is myriads of tiny molecules, or atoms, i.e., micro-entities enacting their micro-powers in a temporally structured order.[7]

In everyday life we are clearly committed to the existence of macroscopic objects, even bigger ones than sugar cubes, such as tables, trees and people. And we commonly ascribe dispositions to these, such as the dispositions to lose the paint, to grow and to fall in love with other people. Following Williams's approach, we have to reconceptualise what occurs to us as genuine parts of reality as derivative entities that supervene upon sequences of manifestations of the micro-powers of

[6]Macroscopic objects are "pseudo objects", as (Williams 2019, ch. 10.2, p. 232) puts it, "as they are complex states of affairs". Being "identical to the simple states of affairs, properly arranged", they "can be reduced without remainder" (Williams 2019, ch. 10.2, p. 228.).

[7]Macroscopic powers exist only linguistically. Raising the question, "what should [...] we say about the typical powers we ascribe to mid-sized objects that result in much longer processes?", Williams explains: "the answer is that these powers will not be real powers at all. They will be powers in name only: rough-and-ready accounts of how things *tend* to occur in the world" (Williams 2014, p. 366; emphasis in the original; see also Williams 2019, ch. 6.4, p. 143). Only "atomic (fundamental) states of affairs are property bearers" and, hence, bearers of powers (Williams 2019, ch. 10.2, p. 229). However, Williams further argues that complex states of affairs, while not having properties, are still "able to satisfy predicates" (Williams 2019, ch. 10.2, p. 226) and thus can truthfully be ascribed 'dispositions' which, as "causally inefficacious predications" (Williams 2019, ch. 10.2, p. 232), are sharply to be distinguished from 'powers'.

micro-entities.[8] Causal processes, such as losing paint, growing and falling in love, are not stories about tables, trees and people and the manifestation of their particular dispositions and powers. They rather become a piece of physics-metaphysics, i.e., something that can be understood only by resorting to the most fundamental level of reality which supposedly is best described by physicists and physics-inspired metaphysicians.

It is doubtful that physicalist reductionism is a good choice for a theory like dispositionalism that gains much of its appeal from its foothold in everyday experience. We experience ourselves as being disposed to certain types of actions and reactions, as agents capable of exerting their causal powers on their surroundings, thus influencing or altering the course of things. This experience, sitting at the core of our self-understanding, is among those things that make many suspicious about Humean accounts of causation as regular succession of otherwise disconnected events.[9] And it also supports the view that powers and dispositions observed at higher-levels of reality cannot, at least not always and necessarily, be reduced to lower-level phenomena or even to phenomena to be found at a hypothetical fundamental level. Mumford and Anjum, defending their commitment to macro-level powers, offer the example of a kiss:

> a kiss to the cheek can have the power to make the recipient blush. Energy may be transferred when the cheek is kissed. The kiss depresses the cheek. But it is not the energy transfer – the cheek depression – that causes the blush. You can depress your own cheek with your finger and it never makes you blush. It is *that it is a kiss* that causes the blush: a high-level, complex matter involving psychological, biological and sociological factors. Perhaps not all of those factors are reducible to physics. But, if they are not, causal dispositionalism can just ascribe the power directly to the kiss and have some powers that are only ascribable to macro-level phenomena. (Mumford and Anjum 2011, p. 102f.; emphasis in the original)

While one may be more sceptical about hypothetical macroscopic dispositions like the disposition to lose paint, dispositionalists are well-advised to take seriously macroscopic dispositions like the disposition to grow or to fall in love. However, this is not to deny that they can also decide otherwise. The decision for or against what one could call microscopism – the view that only the micro-powers of micro-entities are real – does not bear upon the respective theory's status of being a dispositionalist theory, while it does bear upon the way causation is conceptualised

---

[8]"It follows that some true power ascriptions – stated *ceteris paribus* – capture the nature of certain objects, and are handy in science and generally useful for navigating the world around us, but do not pick out the powers that do the work. This is not to deny that dried out hunks of clay are brittle, but what makes it true that such a hunk is brittle is not its possessing a power of brittleness, but that it has some constellation of powers that (if allowed to produce their manifestations through sufficiently many iterations without being disturbed) would give rise to a series of constellations and manifestations on which a process like *crumbling* would supervene" (Williams, Chap. 6, this volume, p. 80; emphases in the original; see also Williams 2019, ch. 6.3, p. 134).

[9]For defences of causal realism within a dispositionalist framework appealing to bodily perceptions of causation see Mumford and Anjum 2011, ch. 9, and Schrenk 2014. For the general implications of dispositionalism for action theory, see my co-edited volume Spann [née Meincke] and Wehinger 2014.

within a dispositionalist framework. We thus can take from Williams the point that the idea of a simultaneity of cause and effect indeed makes more sense if we are not microscopists. To offer another example: my playing the piano is a causal process manifesting my ability to play the piano and involving, apart from me, the piano in my lounge; and this manifestation process lasts as long as the piano and I are co-present in my lounge and I decide to exercise my ability to play the piano.[10] The alternative story told by the microscopist, on the other hand, will – convincingly or not[11] – replace this simultaneity with a temporal succession of neatly separated manifestations of micro-powers, whatever the details (it may, e.g., appeal to cascades of firings of neurons).

Surely, this latter story about causation, despite having powers in it, looks much more Humean than the former: it operates on the assumption prominently invoked by Hume "that causation is a relation that relates two distinct events, objects, or existences" (Anjum and Mumford 2018a, p. 61). This orthodox assumption "immediately suggests a disunited view of the world: that it consists in a succession of distinct, wholly discrete events" (ibid.). Causation, on this view, enters the scene as some mysterious extra-element that *ex post* binds together things that could equally exist without being so connected (Mumford and Anjum 2011, p. 106). Mumford and Anjum doubt that any account of causation presupposing the two-event model could ever be successful:

> if one begins from the idea that causation has to be a relation between two discrete and 'static' events, then it may already be impossible to formulate a satisfactory theory of what causation is and how it works. [...] The assumption splits the world asunder into distinct, self-contained fragments, and then tries to find a relation that would stick them all back together again. It is not clear that any relation can do this job. Even if it could, would the world with which we are left – a Frankenstein-world of stitched together pieces – have all the pieces we require? (Anjum and Mumford 2018a, p. 61f.).

Mumford and Anjum deny the completeness of such a 'Frankenstein-world'. What they think would be missing from it is continuous change, i.e., change that "occurs in a smooth and gradual processual way" (Anjum and Mumford 2018a, p. 61). Acknowledging such continuous change is needed, they insist, for an accurate understanding of causation and, that is, for grasping its inherently dynamic character, which takes us to the question of dynamism.

---

[10]Simultaneity seems generally to hold for the class of powers which Aristotle called rational abilities. However, it does not seem to be true for these powers that they, as Mumford and Anjum's model of causation also requires, 'merge into and become' their effects, at least not literally as in the case of the 'sweet solution'. I'm not turning into a piano while playing the piano!

[11]It is hard to see how any microscopist analysis of my playing, say, the Moonlight sonata on the piano could ever appear to be convincing. Consider my brother Ben who is able to press individually any key I am pressing while playing the Moonlight sonata. Clearly, he thereby does not possess the ability to play the Moonlight sonata. Being able to press keys individually is not the same as being able to play the piano! As Mumford puts it: "you can play the piano only by being able to manifest a process as a whole, and the power has to be for that process" (personal conversation, September 2018).

7.4.2 Dynamism

According to the 'dynamic theory of causation' put forward by Mumford and Anjum, causation is a continuous process that "does not break down into changeless parts" (Anjum and Mumford 2018a, p. 72). This they take to be a corollary of causal dispositionalism: "if one prefers a metaphysics of causal powers to a Humean mosaic of distinct existences, one should have a more dynamic and process-like view of things" (Anjum and Mumford 2018a, p. 63). Indeed, they claim, "there is a close connection between dispositional approaches to causation and process metaphysics, though this has not always been highlighted", due to the broadly Aristotelian orientation of those approaches (Anjum and Mumford 2018a, p. 65). Williams, however, denies such a connection between dispositionalism and process ontology and that is, as Mumford puts it, that "dynamic powers and processes are fundamentally real in the dispositional ontology" (Mumford 2009, p. 230). Williams declares: "I do not think what Mumford says about the dispositional ontology is true; at best this expresses his own view, not that of powers theorists more generally" (Williams 2017, p. 158).

Recall Mumford's speculations about the possibility of a mixed view that combines perdurantism with powers. Mumford conjectured that such a view would adopt so-called "'static powers' which could be instantiated at an unchanging instant" (Mumford 2009, p. 232) so as to resolve the apparent mismatch between dynamic powers and static temporal parts. While Mumford dismisses these hypothetical powers as "unfathomable" (ibid.), Williams explicitly embraces them: his powers for persistence are 'static' in the double sense that they do fit into static temporal parts and that they are powers not for change but for maintaining the status quo, i.e., for bringing about stability (Williams 2005, 2017, p. 146 f., 2019, ch. 7.2). However, Williams emphasises that bringing about stability is an activity, and so it seems one can reject dynamism without thereby being committed to what Brian Ellis called the "dead world of mechanism": a world in which things lack intrinsic activity and which therefore is a world of mere quantity containing only mechanical motion (Ellis 2002, p. 60).

Let's call this hypothetical position staticism. Staticism is opposed to dynamism, which for Mumford and Anjum seems to entail at least the following seven claims: (i) process and change are ontologically primary; (ii) process and change are causally produced by active and dynamic powers; (iii) process and change are essentially extended in time; (iv) each portion of a process is itself a process containing change; (v) change is everywhere, though not always observable as when two interacting powers cancel each other out or bring about an apparently static effect[12]; (vi) any stability thus is the result of process and change; (vii) any (temporal) parts of processes are abstractions rather than ontologically real entities.

[12]Mumford and Anjum aim to capture effects like these (e.g., two books leaning against each other) by modelling dispositional causes as vectors, see Mumford and Anjum (2011), ch. 2. For a critique of this model, see Pechlivanidi and Psillos (Chap. 9, this volume).

Staticism, on the other hand, (i) denies the ontological priority of change, granting priority instead to changeless temporal parts of domino-like processes. However, by conceptualising these temporal parts as adamantine processes of power manifestation, there is a sense in which staticism regards process as elementary: the particular sense of putative short-lived, non-interruptible processes which, according to Williams, do not incorporate any change, i.e., acquisition or loss of properties.[13] Only if put together, adamantine processes may generate change, but not necessarily so, as the case of persistence shows. Accordingly, (ii) staticism agrees that there are active powers that produce process and change, while insisting that there are also active powers whose activity consists in bringing about the *opposite of change*, i.e., stasis or stability, which is why these powers are called static powers.[14] It therefore, according to staticism, is wrong to equate process and change.

Furthermore, staticism, (iii), denies that processes are essentially extended in time by forming non-instantaneous adamantine wholes. If processes are adamantine, they are instantaneous (or close to instantaneous); if they are not instantaneous (or close to instantaneous), then they are domino-like; and only the latter kind of process can manifest change. Staticism, thereby, (iv), denies that each portion of a process is itself a process that contains change, construing change and (domino-like) process from changeless static existences instead. This means that staticism, (v), denies the ubiquity of change assumed by dynamism. Where we can't see any change, there is likely to be none. According to staticism, (vi), stability does not result from process and change in the sense of being brought about by dynamic powers; it is rather brought about by static powers and tantamount to the absence of change. Finally, (vii), unlike dynamism, staticism regards the temporal parts of protracted processes as ontologically real entities providing the cleavages guaranteeing interruptibility.

Dynamism and staticism thus disagree on the nature of process and change, but they share the assumption that process and change – whatever they are – are produced by active powers. Moreover, staticism endeavours to account for an aspect of reality apparently overlooked (if not dogmatically excluded) by dispositionalist accounts that subscribe to dynamism: stability, that is, the absence of change. Without assuming static powers, such as persistence powers,[15] Williams thinks, one would be committed to assuming that the absence of change in the world amounts to an absence of causation, which not only goes against the core tenets of dispositionalism but would also be very odd indeed. It would mean that the universe would at times stand still, followed by all-of-a-sudden appearances of causation (Williams 2005, 2019, ch. 7.2). Powers for persistence, in particular, are needed to

[13]Change "must involve bringing about a state of affairs that is clearly different from that which obtained prior to the manifestation of the disposition" (Williams 2005, p. 304).

[14]Williams claims that we tend to overlook these static powers because our survival more strongly depends on knowing dynamic powers, and science is primarily interested in explaining change (Williams 2005, p. 307 f.).

[15]Williams (2005) distinguishes between three kinds of static powers: static powers for internal stability (which are powers for persistence), static powers for external stability and static powers involving threshold conditions.

account for the 'pluckiness' of material objects, i.e., their tendency to continue to exist "even when faced with difficult scenarios" (Williams 2019, ch. 8.2, p. 177). According to Williams, the 'problem of pluck' is the real problem of persistence which must be distinguished from the problem of change with which it is commonly conflated. Considering a case of persistence without changes, the relevant question to ask is not how an object *a* "can be the same throughout changes" but rather: "how can *a* exist at two distinct times at all?" (Williams 2019, ch. 8.2, p. 178). Williams complains that this question has been largely ignored by metaphysicians so far (Williams 2017, p. 146, Williams 2019, ch. 8.2; see also Simons unpublished for similar considerations).

There are obvious reasons for radical anti-Humeans to dislike the ontological picture sketched by Williams: to them, this picture looks like Humeanism with powers as add-ons, i.e., as something that enhances the picture without substantially revising it. Powers serve the function of connecting the disconnected facts of the Humean mosaic, which, as we have seen, prompts the question of why we should start with such a mosaic in the first place. However, does this suffice for disqualifying Williams's powers metaphysic, including its perdurantist account of persistence, from being what it declares to be: a version of dispositionalism? Preferences as to the overall ontological picture of reality are one story; what makes the ontological picture one subscribes to dispositionalist is another. We have to acknowledge the possibility of positions within the conceptual space of dispositionalism that do not discard process and change altogether but take a more moderate stance on their existence and ontological status.

Powers, it turns out, bear fewer constraints upon general ontology than commonly assumed. Instead, they can be modelled in accordance with preferences and convictions held for reasons independent of the commitment to dispositionalism. As long as there is no generally accepted rule about how much dynamism – activity, change and process – is needed for dispositionalism, we are well-advised to allow in the dispositionalist camp any theory that endorses realism about dispositional properties and causation and that incorporates at least some dynamist elements such as those to be found in Williams's staticist dispositionalism. This is, of course, not to say that staticist dispositionalism is more attractive than dynamist dispositionalism. It isn't, I contend. But this is a different matter to which I shall turn now.

7.5 Powerful Persistence and Process

As we have seen, Williams rejects Mumford and Anjum's concept of causal processes on the grounds that the interruptibility commonly attributed to protracted processes requires these to be domino-like rather than adamantine. There are at least three problems with this argument.

First, the argument presupposes that Mumford and Anjum take the processes in which powers manifest themselves to be adamantine in the sense of 'not being interruptible'. Strikingly, this presupposition is false. When Mumford and Anjum (2011, p. 116) state: "change is undergone throughout the process, which means it is to be found in any part of it, and it thus cannot be broken down into a string of changeless parts", they highlight the continuous change involved in any such process. Nothing in this characterisation permits Williams's inference that processes in Mumford's understanding are "impenetrable",[16] which, of course, would be absurd and would make in turn the domino-model of process appear to be the only viable alternative. In fact, interruptibility is germane to Mumford and Anjum's assumption of a genuine dispositional modality lying between necessity and contingency (Mumford and Anjum 2011, ch. 3 and ch. 8; Anjum and Mumford 2018b),[17] without this by any means affecting their ontological claim that processes are not dissectible into static, unchanging parts. For a manifestation process to be interrupted, it simply needs the intervening of other powers. Mumford's processes are not 'adamantine'.[18]

Second, as demonstrated by Mumford and Anjum's theory, it is anything but obvious that the interruptibility of a process "requires that there are natural breaks in that process", that is, "natural cleavages into which metaphorical spanners can be inserted" (Williams 2017, p. 159). Williams argues that only because a protracted process does not come in one piece but unfolds stepwise is it possible for new powers to act upon the unfolding process and to possibly divert or disrupt it, by modifying or preventing the manifestation of the powers involved in the process's latest step.[19] But this presupposes without argument (i) that powers work at the micro-level only and (ii) that powers can be possessed exclusively by some sort of static thing, such as short-lived, possibly instantaneous temporal parts (analysed as complex states of affairs). That is, Williams's argument from interruptibility for the domino-model of process, rather than providing independent support for powerful perdurance, in fact can appear plausible only if one is already convinced of the corresponding overall ontological picture – a picture characterised by what I have called microscopism and staticism.

Third, and most importantly, Mumford and Anjum convincingly argue that, in any case, we *should not* have any cleavages in our ontology. For if there were such

[16]"The second model of processes, the 'adamantine' model, sees them as continuous and impenetrable. To split them up is impossible, as the process would thereby cease to be a member of the same process-type, or cease to be natural in the right sort of way, or fail to be properly understandable" (Williams 2017, p. 157).

[17]Williams rejects this assumption, insisting that the (micro-) powers at the fundamental level manifest with necessity whereas interruptibility is a higher-level phenomenon exhibited only by domino-like processes: "Manifestations arise as a matter of metaphysical necessity, but the processes that supervene on sequences of manifestations do not" (Williams 2014, p. 366).

[18]The concept of a temporally extended 'adamantine' and, that is, of a non-interruptible temporally extended process looks indeed like a strawman to me. Who would ever endorse such a concept? The only philosopher I know of who endorses an adamantine concept of process at all is Williams himself with regard to the fundamental level of the manifestations of microscopic powers.

[19]Compare also Williams's example of the Turing machine in Williams (Chap. 6, this volume, p. 75).

cleavages, e.g., in processes, it would become mysterious how to get from one temporal part to the next.[20] Williams, in reaction, insists that this is what powers are for: linking numerically different existences, such as temporal parts. However, the fact that the activity of powers is confined to single temporal parts raises doubts if such 'linking' really takes place. Each temporal part is said to contain a power which is the cause of the following temporal part which manifests it. But how is it that the cause produces the effect given that there is a gap between the two?

Reconsider the diagram above (Fig. 7.1). The perpendicular lines mark the boundaries between the single temporal parts while the curly brackets indicate the adamantine short processes of manifestation. The arrows stand for the causation that is supposed to link the temporal parts. The position of the arrows between the temporal parts is misleading, though. All the manifestation of powers which gives rise to new temporal parts happens *within* the individual temporal parts. The individual temporal parts *are* the manifestations. Williams conceptualises these as short-lived adamantine (i.e., non-interruptible) processes so as to capture the activity of powers. However, this does not change the fact that once such an adamantine manifestation process is completed (and keep in mind: we are talking here about instantaneous or nearly instantaneous processes), nothing takes us to the next adamantine manifestation process. There is, so to speak, no ontological bridge over which we could walk to get from one temporal part to the next; there is just a gap.

Despite Williams's rhetoric, then, there are no *links* between temporal parts properly speaking.[21] And these temporal parts never make up a causal *chain* properly speaking. To wit, the elements of a chain are interlocked with one another (and only therefore are called 'links'), whereas Williams's temporal parts are merely juxtaposed (and thus do not deserve being called 'links'). While Williams's account, unlike the standard Humean two-events model, may indeed *not* imply that "the two events in question, or two just like them, might also exist without the causal relation connecting them", which would "invite[ ] causal scepticism" (Mumford and Anjum 2011, p. 106), the connection between the two events still remains mysterious. It is not the case that causation "stretches across the gap and ties them together" (ibid.); the process of causation rather comes to an abrupt end just before the effect – a new state of affairs – occurs. The only way to cross the gap, if there is any, is to jump.

I find it hard to believe that reality should behave like a sputtering engine or, to use another image, like an athlete who takes a run-up for a jump that miraculously then happens outside time, before taking a new run-up immediately after, followed by another miraculous jump, and so on. This picture, qua containing some rudimentary form of activity, may be better than the common (neo-) Humeans'

---

[20]Anjum and Mumford (2018a, p. 72 f.) accordingly defend the claim that time is dense rather than discrete.

[21]"I think that there's a temporal part, and then there's another. The 'causal magic' (the acting of the powers of the first in producing the second) are best thought of as acting between. Of course, that's misleading. There's no time involved. There's really just one stage and then the next. But the arrows show that that's where the causal action is, so to speak. If some mighty being ended the world at TP1, TP1 wouldn't bring anything about" (N. Williams, personal conversation, July 2013).

"Frankenstein-world of stitched together pieces" (Anjum and Mumford 2018a, p. 61 f.), but I agree with Mumford and Anjum that something important is still missing here: continuous change – or even change altogether.

Williams is aware of this type of objection against perdurantist accounts of persistence, which he initially dismisses out of hand. Having explained that temporal parts are static, i.e., "incapable of change", and that they "exhibit change collectively" only (Williams 2017, p. 155 f.), he comments in a footnote:

> this has led to the objection that perduring objects do not in fact change [...]. Though I recognize that intuitions about these things can differ, I cannot help but treat the no-change objection as anything but an uncharitable understanding of change. The perduring object has different properties at different times – this is change enough. (Williams 2017, p. 156)

A little later he returns to the subject, though, asking the reader to consider the domino model of process, which grounds the perdurantist analysis of persistence, as a "fine-grained perspective" that should not alter "our understanding of the process as a whole" (Williams 2017, p. 157): the process does not "lose something of its dynamic and developing nature" (ibid., quoting Mumford 2009, p. 228) by understanding it "as dissectible into ever-so-many parts", just as "we do not fail to see the forest for the trees when we learn that forests are little more than a collection of trees" (Williams 2017, p. 157). This, Williams emphasises, remains true even if we assume that "processes depend for their existence on 'ontologically prior' temporal parts" (Williams 2017, p. 157 f.).

There is an obvious difficulty with the analogy used by Williams: a forest, considered as a collection of trees, is no process. Hence, one might argue, the image of the forest is of little or no help when it comes to understanding the dynamicity of a process. If a persisting object was indeed like a forest in Williams's sense,[22] then this would be to say that it is a collection of discrete static existences – but not a process! Indeed, I contend, this is exactly what is the case with persisting objects as conceived by perdurantism. They are no processes proper, and they do not change. They are collections of discrete static existences – full stop.

Williams recognises the existence of people like me who, despite their best efforts, find themselves unable to follow him here. He remarks:

> I suspect that at least some of those who remain unconvinced simply refuse to see the dynamic process that powers produce as fitting with the static temporal parts the perdurantist supports. These might well be the same folk who conclude from Zeno's paradox of the arrow

[22]The qualification "in Williams's sense" is crucial here since to characterise a forest as a "collection of trees" is, of course, to completely misunderstand what a forest is. As we can learn from forest ecologists, forests are complex ecosystems consisting of numerous mutually dependent species of organisms, including trees. Trees, rather than being static, distinct things, rely on one another as well as on other species for their survival by way of continuous interaction, something Whitehead (1925, p. 206) acknowledged when calling the forest "the triumph of the organisation of mutually dependent species". Ontologically, thus, forests are much better understood as dynamic networks of interdependent processes, and *in this sense*, the process ontologist maintains, they may indeed serve as a model of persisting entities.

> that motion simply cannot be understood in terms of ever-so-many motionless parts occupying different places at different times. (Williams 2017, p. 163)

Very well. And exactly because the latter speculation sounds right to me, I would like to clarify that I don't see any proper dynamic process in powerful perdurance at all. Williams tries to sneak in some dynamicity by reconceptualising temporal parts as adamantine processes of power manifestation. However, apart from these being rather dubious entities, they remain discrete existences and, qua unchanging, static in themselves. Indeed, there is an odd tension between the supposed (adamantine) processual character of temporal parts and their supposed static character. Processes, by definition, are not static but dynamic. Williams refers to the absence of change in a temporal part when characterising it as static, i.e., to the fact that the temporal part does not lose or gain a property.[23] But this is just another oddity. Can we have process without change? Certainly, the world changes when dispositions get manifested. It makes a huge difference to me whether I merely have a disposition to breast cancer or whether this disposition gets manifested. And I would expect that the manifesting of that disposition involves all sorts of changes along the way. Admittedly, a disposition to breast cancer is probably best analysed as a complex macroscopic disposition. But it is hard to see why the (hypothetical) microscopic case should be any different. It is hard to think of power manifestation as not being a continuous process involving continuous change.

Henri Bergson has convincingly argued that the illusion that one could compose processes from static temporal parts stems from our tendency to spatialise time. If you draw a line tracking the movement, say, of your hand, then, of course, this line is infinitely divisible. But the movement of your hand is not, which is most compellingly proved by the fact that your hand, by moving, reaches a different position in a finite amount of time. According to Bergson, dissecting process into static things is a mental operation, or abstraction, which serves certain practical functions (including scientific ones) but which cannot claim to depict reality (Bergson 2004, ch. 4, and Bergson 2010, ch. 5 and ch. 6). Anjum and Mumford touch on this point when they explain:

> processes have no motionless parts. Indeed, in no real sense does a process have parts at all; for they can be formed only through the abstraction of a viewer who considers the process. The process is, in reality, an indivisible unity. (Anjum and Mumford 2018a, p. 71)[24]

---

[23] "[T]he mere change in status of a disposition from non-manifested to manifested will not count as an appropriate change in this sense [. . .] as no property is actually lost or gained by the object in this situation" (Williams 2005, p. 321).

[24] See also Mumford and Anjum (2011), p. 121: "Causation comes in temporally extended wholes rather than as constructions from changeless discreta".

Temporal parts, in other words, are no real existences; they are abstractions. And as Bergson makes clear, one can build an abstraction from the real, but one cannot build the real from an abstraction from the real.[25]

So there are only these two options then, as far as our basic ontology is concerned: either we are what I call thing ontologists: philosophers who give the ontological priority to *things*, i.e., to entities for the identity of which change is not essential. Or we are process ontologists who take *processes* to be ontologically primary, i.e., entities for the identity of which change is essential (Meincke 2018b, 2019a, b, c, forthcoming). Note that it follows from these definitions of 'thing' and 'process' that Williams's processes – both adamantine and domino-like – are processes in name only. They don't change – hence, they are things: tiny, short-lived things or collections of such tiny, short-lived things. Perdurantism, with or without powers, is a version of thing ontology on account of its broadly Humean orientation (Meincke 2019b, c).[26]

What matters in the present context is that depending on which option we choose – thing ontology or process ontology – our dispositionalism, including our dispositionalist account of persistence, will look different. And equally, depending on which option we choose, different versions of powerful persistence will look differently convincing or attractive to us. The reasons why powerful perdurance does not appeal to me are largely the same that draw me away from thing ontology towards process ontology. This is what comes out from Mumford's argument against powerful perdurance: if you are a process ontologist, or, at least, if you have strong process ontological leanings, then powerful perdurance is not for you.

Some people may say: here philosophical dispute ends; one cannot argue about intuitions. The opposite is true. To start talking about our most fundamental ontological commitments is exactly where things become interesting and true philosophy begins.

Process ontology has a lot going for it, evident in particular from the recent call in philosophy of biology for acknowledging the processual nature of organisms (e.g., Nicholson and Dupré 2018; Jaeger and Monk 2015; Dupré 2012; Bickhard 2011). Organisms are dynamical systems which for their existence depend on constant interaction with the environment in which they are situated, most fundamentally on a constant exchange of matter and energy with the environment (metabolism).

[25] See, e.g., Bergson (2010), ch. 6, p. 212 f.: "One recognizes the real, the actual, the concrete, by the fact that it is variability itself. One recognizes the element by the fact that it is invariable. And it is invariable by definition, being a schema, a simplified reconstruction, often a mere symbol, in any case, a view taken of the reality that flows. But the mistake is to believe that with these schemas one could recompose the real."

[26] Similarly, Anjum and Mumford explain that the standard model of causation starts from "a discontinuous view of nature, because change is analysed as a mere succession of discrete, self-contained objects, parts, or stages. The perdurantist begins with a metaphysics of static, temporal parts and then tries to explain change through the way in which those parts are arranged. The alternative view is to accept change as basic – something to be found in every segment of a process, no matter how small – and then explain stability [. . .] as an equilibrium created by counterbalancing powers" (Anjum and Mumford 2018a, p. 71).

Persistence, for living beings, is a matter of stability, where stability is exactly not the same as stasis or the absence of change. Instead, stability is contingent upon on the occurrence of certain kinds of changes, including certain kinds of interactions with the environment; it requires active change or even, in higher organisms, agency. If the interaction breaks down, stability breaks down and the organism dies (Meincke 2018a, b, 2019a, b, c, forthcoming). Compare this to the picture of persistence sketched by Williams. Persistence is portrayed as a "perpetuation or conservation" of the status quo (Williams 2019, ch. 9.4, p. 214), brought about by static powers which, as we have seen, are supposed to be unilateral, i.e., not to need any mutual manifestation partners. Williams explicitly denies that persistence is interactive. Persistence becomes a matter of an object's self-reproduction or self-re-invention from one moment to the next,[27] a sort of isolated and scattered inertia tuned with powers. This is clearly not how things work in biology.[28]

From a process perspective, there is no persistence without change. Change is not only, as is commonly assumed, a false-maker of statements of persistence. It is also, and most importantly, a truth-maker of statements of persistence. Dependent on what kind of entity we are concerned with, different sciences are in charge of investigating what kinds of changes are needed for that entity to maintain its stability through time. Identity, synchronic and diachronic, is never a pre-given but always a (more or less hard-won) interactive achievement (Meincke 2018b, 2019a, b, c, forthcoming).

For the same reason it is anything but clear that we would be substantially better off if analysing powerful persistence in terms of endurance rather than perdurance. Enduring entities are standardly taken to be substances. Substances are *things* – just bigger ones than the atomic things regarded as fundamental by the Humean ontology.[29] Aristotle-inspired endurantists will, of course, emphasise that substances are *active* things – things that *do* stuff, namely by means of their various

[27]"[I]n order for the object as a whole to persist, the object stages themselves must have powers to reinvent themselves from one moment to the next. The object is involved in a perpetual state of reproduction: each object stage creating the next, moment after moment" (Williams 2019, ch. 9.4, p. 214).

[28]Williams may want to reply that he doesn't care so much about biology because biology is reducible to physics, and for physics, his picture of persistence is the most plausible one. However, apart from it being doubtful if biology is indeed reducible to physics, it is not clear that physics supports a thing ontological view of reality. According to Bergson, the opposite is the case: "The more [physical science] progresses the more it resolves matter into actions moving through space, into movements dashing back and forth in a constant vibration so that mobility becomes reality itself" (Bergson 2010, ch. 5, p. 175; see also Bergson 2004, ch. 4, pp. 263ff.). Similarly, Rescher (1996, p. 97 f.) argues that "the rise of quantum theory put money in the process philosopher's bank account" insofar as "matter in the small" turns out to be "a collection of fluctuating processes organized into stable structures [. . .] by statistical regularities". In striking parallel with Bergson's interpretation of matter as a continuum of rhythmic vibrations, some physicists have recently started promoting what they call "eurhythmic physics" which analyses physical entities as maximally stabilised systems of interactions (Croca 2016). I am grateful to Leo Caves for pointing me towards Croca's work.

[29]For a more detailed discussion of the distinction between substance ontology and Humean ontology as two types of thing ontology, see Meincke 2018b, 2019b, c, forthcoming.

powers. From a (neo-) Aristotelian viewpoint, hence, there seems to be a natural and intimate link between dispositionalism and dynamism. However, this sort of dynamism (as defended, e.g., by Groff 2013, Chap. 3, this volume) must not be confused with the serious dynamism of the process ontologists. The hallmark of thing ontology is that process and change are made parasitic upon static things rather than coming first. This is the case also in (Aristotelian) endurantism.[30] Substances are thought to be unchanging bearers of change; what they *do* is processual, but not what they *are*. There is no way to investigate what it takes for a substance to persist because substance identity is primitive. Substances are ontologically independent particulars, not in need of interaction with the environment. They can happily sit there without doing anything. In contrast, a process account of persistence, as I envisage it, would analyse persistence as stability brought about by complex interactions of powerful processes (Meincke 2018b, 2019b, c, forthcoming).

7.6 Conclusions

In this chapter I have examined Mumford's thesis that dispositionalists ought to be endurantists, or at least ought not to be perdurantists, by way of confronting this thesis with Williams's theory of powerful perdurance. It should have become clear that we have little reason to think that Mumford is right. Apart from not being given any positive arguments for why dispositionalism and endurantism should be a good match (strong claim **C**), the negative arguments we are provided with for why dispositionalism and perdurantism are no good match (weaker claim $\mathbf{C}^*$) rest on metaphysical assumptions that go beyond the core tenets of dispositionalism. As illustrated by Williams's alternative approach, there is some leeway as to how to construe our concept of powers and, consequently, our concepts of causation, process and persistence. Which of the possible variants we choose depends on which position we adopt with respect to the alternatives of (what I have called) macroscopism versus microscopism and dynamism versus staticism, and more generally on where we place ourselves on the spectrum between (more) Humean and (more) anti-Humean overall perspectives. Williams's account is located closer towards the former end, whereas Mumford's is located at the latter. Insofar as both appear to be perfectly legitimate and respectable versions of dispositionalism or powers metaphysics, Mumford's argument against powerful perdurance is defeated.

Mumford's argument remains instructive in other ways, though. It points towards a deeper antagonism – an ontological antagonism which is even more

[30]For a detailed analysis of endurantist accounts of persistence, see especially Meincke 2019b; for some reflections on the relative advantages of Aristotelian endurantism, and Aristotelian substance ontology as such, over other versions of endurantism and substance ontology, see Meincke 2019b, c.

fundamental than the antagonism between Humeanism and anti-Humeanism: ultimately, which version of dispositionalism we prefer will depend on whether we have thing ontological or rather process ontological leanings. The concept that most significantly bears witness of this difference is the concept of process. If you are a thing ontologist, you will make your change contingent upon changeless entities – 'things' – whose identity you take for granted; and, at least as a perdurantist thing ontologist, you will think that processes are just ordered collections of static things. If you, however, are a process ontologist, you will turn the tables and claim that there is no identity, and no persistence, without change, and that processes cannot be made up from unchanging, static things but are rather irreducibly and continuously changing as they move on. Thus, while we have to concede to Williams that the close connection between dispositionalism and process metaphysics championed by Mumford falls into the category of personal preferences, it is also true that the conviction that there are ontologically primitive, essentially dynamic persisting processes provides a strong motivation for rejecting the picture of powerful perdurance sketched by Williams. In other words: it is, contrary to what Mumford claims, not inconsistent to combine powers with perdurance; but it is open to debate whether the resulting position is convincing. From a process ontological viewpoint, this is clearly not the case.

This outcome of the investigation has important implications for future research on both powers and persistence. First, reflection on the ontological commitments of dispositionalism has to be put in the context of wider metaphysical orientations and frameworks. There is not 'the' powers metaphysics; there are different interpretations of what a powers metaphysics is or ought to be – shaped by more general views on reality. To put it as a slogan: it's not that powers dictate what metaphysics you have, but rather the metaphysics you have dictates what your powers look like and, hence, what your powerful persistence looks like.[31] Second, the debate in metaphysics on persistence should come to acknowledge the significance of concepts of process for accounts of persistence. Process, properly understood, offers a promising starting point for making sense of the copresence of identity and change characteristic of persistence as we find it in our finite world. Accordingly, third, more investigation is needed as to the possibility of fruitful links between dispositionalism and process ontology. It is not clear that these two go well together, given that dispositionalists tend to be committed to substances or (at least) particulars as bearers of dispositions. However, nothing in principle speaks against ascribing powers to processes. Substances, after all, may need to be reconceptualised as stable patterns of powerful processes. Accordingly, my take on powerful persistence is that dispositionalists ought to endorse neither perdurantism nor endurantism. They rather ought to be process ontologists and endorse a process account of persistence. But

[31]This accords with Heil's thesis that powers are part of an overall ontological 'package deal', see Heil (Chap. 2, this volume). However, Heil and I disagree on what this overall ontology should be: Heil endorses (a version of) substance ontology; I favour process ontology.

this is, as I have argued here, not a necessity flowing from dispositionalism as such. It is rather what is recommendable for a version of dispositionalism that takes seriously the dynamicity of reality.

Acknowledgements This chapter brings together research I conducted at the University of Innsbruck, Austria, under the auspices of the research project "Powers and the Identity of Agents" (PI: E. Runggaldier), funded by Austrian Science Fund, and at the University of Exeter, UK, under the auspices of the research project "A Process Ontology for Biology" (PI: J. Dupré), funded by the European Research Council. I completed the final version of the chapter while working on the research project "Better Understanding the Metaphysics of Pregnancy" (PI: E. Kingma) at the University of Southampton, UK, also funded by the European Research Council. Many thanks to all institutions and funders. The chapter has benefitted from three public presentations, given on 1st August 2013 in Innsbruck, Austria, at the conference "The Ontological Commitments of Dispositionalism", organised by me and D. Wehinger as part of the project "Powers and the Identity of Agents"; on 20th May 2015 in Exeter, UK, as part of the public lecture series at the Centre for the Study of Life Sciences (Egenis) of the University of Exeter; and on 21st September 2018 in Cologne, Germany, at the workshop "Change and Change-Makers. New Perspectives on the Problem of Persistence", organised by the Society for Philosophy of Time (S.P.O.T.) in association with the 10th Congress of the German Society for Analytic Philosophy (GAP). I am grateful to the audiences in Innsbruck, Exeter and Cologne for inspiring discussions. Last but not least, I am indebted to Neil E. Williams for his efforts to help me understand his view, to Stephen Mumford for very useful comments on an earlier draft of this chapter and to an anonymous reviewer for their helpful feedback.

References

Anjum, Rani L., and Stephen Mumford. 2018a. Dispositionalism. A Dynamic Theory of Causation. In *Everything Flows. Towards a Processual Philosophy of Biology*, ed. D. J. Nicholson and J. Dupré, 61–75. Oxford: Oxford University Press.

———. 2018b. *What Tends To Be: The Philosophy of Dispositional Modality*. London/New York: Routledge.

Bergson, Henri. 2004. *Matter and Memory*. Trans. N. M. Paul and W. S. Palmer. Mineola, NY: Dover Philosophical Classics.

———. 2010. *The Creative Mind*. Trans. M. L. Andison. Whitefish, MT: Kessinger Legacy Reprints.

Bickhard, Mark. 2011. Systems and Process Metaphysics. In *Handbook of Philosophy of Science. Philosophy of Complex Systems, Vol.* 10, ed. C. Hooker, 91–104. Amsterdam: Elsevier.

Croca, José N. R. 2016. The Unity of Physis. *Philosophica (Lisboa)* 47: 141–153.

Dumsday, Travis. 2019. *Dispositionalism and the Metaphysics of Science*. Cambridge: Cambridge University Press.

Dupré, John. 2012. *Processes of Life. Essays in the Philosophy of Biology*. Oxford: Oxford University Press.

Ellis, Brian. 2001. *Scientific Realism*. Cambridge: Cambridge University Press.

———. 2002. *The Philosophy of Nature: A Guide to the New Essentialism*. Montreal/Kingston: McGill-Queen's University Press.

Groff, Ruth. 2013. Whose Powers? Which Agency? In *Powers and Capacities in Philosophy. The New Aristotelianism*, ed. R. Groff and J. Greco, 207–227. New York: Routledge.

———. This volume. What Does the Doing? On Powers, Things, and Powerful Things. In *Dispositionalism. Perspectives from Metaphysics and the Philosophy of Science (Synthese Library* 417*)*, ed. A. S. Meincke, 27–40. Cham: Springer.

Hawley, Katherine. 2001. *How Things Persist*. Oxford: Oxford University Press.
Heil, John. This volume. Ontology of Powers. In *Dispositionalism. Perspectives from Metaphysics and the Philosophy of Science (Synthese Library* 417*)*, ed. A. S. Meincke, 13–26. Cham: Springer.
Jaeger, Johannes, and Nick Monk. 2015. Everything Flows: A Process Perspective on Life. *EMBO Reports* 16 (9): 1064–1067.
Johnson, William E. 1964. *Logic. Part III: The Logical Foundations of Science*. New York: Dover.
Lewis, D.K. 1986. *On the Plurality of Worlds*. Oxford: Blackwell.
Meincke, Anne Sophie. 2018a. Bio-Agency and the Possibility of Artificial Agents. In *Philosophy of Science. Between the Natural Sciences, the Social Sciences, and the Humanities (European Studies in Philosophy of Science* 9*)*, ed. A. Christian, D. Hommen, N. Retzlaff, and G. Schurz, 65–93. Cham: Springer.
———. 2018b. Persons as Biological Processes. A Bio-Processual Way Out of the Personal Identity Dilemma. In *Everything Flows. Towards a Processual Philosophy of Biology*, ed. D. J. Nicholson and J. Dupré, 357–378. Oxford: Oxford University Press.
———. 2019a. Autopoiesis, Biological Autonomy and the Process View of Life. *European Journal of Philosophy* 9 (5). https://doi.org/10.1007/s13194-018-0228-2.
———. 2019b. The Disappearance of Change. Towards a Process Account of Persistence. *International Journal of Philosophical Studies* 27 (1): 12–30. https://doi.org/10.1080/09672559.2018.1548634.
———. 2019c. Human Persons – A Process View. In *Was sind und wie existieren Personen?* ed. J. Noller, 57–80. Münster: Mentis.
———. Forthcoming. Processual Animalism. In *Biological Identity. Perspectives from Metaphysics and the Philosophy of Biology (History and Philosophy of Biology)*, ed. A. S. Meincke and J. Dupré. London: Routledge.
Mumford, Stephen. 2009. Powers and Persistence. In *Unity and Time in Metaphysics*, ed. L. Honnefelder, E. Runggaldier, and B. Schick, 223–236. Berlin: de Gruyter.
Mumford, Stephen, and Rani L. Anjum. 2011. *Getting Causes from Powers*. Oxford: Oxford University Press.
Nicholson, Daniel J., and John Dupré, eds. 2018. *Everything Flows. Towards a Processual Philosophy of Biology*. Oxford: Oxford University Press.
Pechlivanidi, Elina, and Stathis Psillos. This volume. What Powers are Not. In *Dispositionalism. Perspectives from Metaphysics and the Philosophy of Science (Synthese Library* 417*)*, ed. A. S. Meincke, 131–149. Cham: Springer.
Rescher, N. 1996. *Process Metaphysics. An Introduction to Process Philosophy*. Albany: State University of New York Press.
Schrenk, Markus. 2014. Die Erfahrung der Widerständigkeit der Welt als Wahrnehmung kausaler Kraft. In *Vermögen und Handlung. Der dispositionale Realismus und unser Selbstverständnis als Handelnde*, ed. A. S. Spann [née Meincke] and D. Wehinger, 23–62. Münster: Mentis.
Simons, Peter. Unpublished. *Keep Going. The Motor of Persistence*. Unpubl. manuscript.
Spann [née Meincke], Anne Sophie, and Daniel Wehinger, eds. 2014. *Vermögen und Handlung. Der dispositionale Realismus und unser Selbstverständnis als Handelnde*. Münster: Mentis.
Whitehead, Alfred N. 1925. *Science and the Modern World*. Cambridge: Cambridge University Press.
Williams, Neil E. 2005. Static and Dynamic Dispositions. *Synthese* 146 (3): 303–324.
———. 2014. Powers: Necessity and Neighborhoods. *American Philosophical Quarterly* 51 (4): 357–371.
———. 2017. Powerful Perdurance: Linking Parts with Powers. In *Causal Powers*, ed. J. D. Jacobs, 139–164. Oxford: Oxford University Press.
———. 2019. *The Powers Metaphysic*. Oxford: Oxford University Press.
———. This volume. What Are Manifestations? In *Dispositionalism. Perspectives from Metaphysics and the Philosophy of Science (Synthese Library* 417*)*, ed. A. S. Meincke, 67–87. Cham: Springer.

Chapter 8
Powers, Probability and Statistics

Rani Lill Anjum and Stephen Mumford

Abstract Regularity is often taken as the starting point of our causal knowledge. But pure constant conjunctions are not what science finds. Even in randomised controlled trials, we do not discover a regular frequency of occurrence of some effect. The dispositionalist is able to explain the evidence of science in terms of the ontology of real causal powers exhibiting an irreducibly tendential nature: less than necessity but more than pure contingency. Much evidence of this kind has to be understood probabilistically and there is a frequentist interpretation in which the facts of frequency of occurrence fix all the facts of probability. However, the dispositionalist has a stronger propensity interpretation of probability at their disposal in which the facts of probability are determined by the individual powers of things. The dispositional approach allows us to make sense of large-scale population data in which different individuals within the same sub-group can have different probabilities of being affected by a cause. On this view, individual propensities can compose to make an overall chance of an effect for a group. But from a starting point of general facts of probability for groups, we cannot decompose those chances back to individuals.

Keywords Causation · Powers · Probability · Propensity · Frequency · Data · Statistics · Randomised controlled trials · Medicine

---

R. L. Anjum (✉)
School of Economics and Business, Norwegian University of Life Sciences (NMBU), Ås, Norway
e-mail: rani.anjum@umb.no

S. Mumford
Department of Philosophy, University of Durham, Durham, UK

School of Economics and Business, Norwegian University of Life Sciences (NMBU), Ås, Norway
e-mail: stephen.mumford@durham.ac.uk

A. S. Meincke (ed.), *Dispositionalism*, Synthese Library 417,
https://doi.org/10.1007/978-3-030-28722-1_8

8.1 Causation and Regularity

David Hume famously connected the notion of cause with regularity (Hume 1739, I, iii, 2, p. 77). His reason for doing so was his rejection of singularism. When he looked at a single instance, all he could see was one 'object' or event followed by another. He couldn't detect that the first event necessitated another or that, without the first, the second would not have happened either. We would never arrive at the idea of the first event causing the second, he concluded, without the experience of repetition. If there are many instances of the first type of event and they are all followed by instances of the second type of event, then we might start to form the idea of the first being a cause of the other. We have experience of many cases of the eating of apples, for example, and they have all been followed by the nourishing of the body.

From this accumulation of repeated instances, we form the idea that apples nourish or, to use explicit causal vocabulary, eating apples causes nourishment. Crucially, however, we would never have arrived at this conclusion if our experience was limited to just one instance of an apple being eaten, followed by the body being nourished. And nor would we have drawn this conclusion just from the inspection of the apple, for reason alone could not show us that the properties of being round, smooth and green would be followed by nourishment. The only way we acquire the idea of causation, Hume concludes, is regular succession or what is known as constant conjunction.

Regularity has remained deeply associated with our causal thinking. The idea is then that where one factor A causes another B, it should do so on all occasions. Both Humeans and anti-Humeans seem to an extent to agree on this. Humeans believe there is nothing more to causation over and above that pattern of succession of events. If A is always followed by B, then nothing more is required for it to be a fact that A causes B, except for the two other conditions Hume imposes: that A occurs before B and that they are spatially contiguous (for a defence of the Humean approach, see Psillos 2002). Anti-Humeans tend to think that these conditions are not enough and that there has to be some necessity that produces the regularity (see for instance Ellis 2001). But, either way, there is agreement that regularity is a part of causation.

This is a view we will challenge, however. The connection between causation and regularity is not quite so clear cut if one accepts the insights of a dispositional theory of causation based on a metaphysics of real powers (Mumford and Anjum 2011a, b). We will argue that a world of tendencies requires a reappraisal of the traditional view for it opens the possibility of causation without constant conjunction. The world of tendencies may still exhibit a degree of regularity, however: enough for us to get by. But, as the notion of degree suggests, this regularity will be less than perfect. We will argue that this is entirely consistent with typical results when we investigate the world scientifically. Setting aside the debates around whether such a view is metaphysically adequate or not, we will be claiming that such a metaphysics is capable of producing a satisfactory epistemology. The way that we have come to understand correlation in the discovery of causes, for instance, is one that allows for the tendential nature of causation. And where we have irreducible probabilities, these are to be understood in terms of propensities rather than relative frequencies. Importantly, however, such tendencies must retain their irreducibly dispositional nature and should not resolve into necessities, which would again require constant conjunction in causes and effects.

8.2 Tendencies

The metaphysics of tendencies offers a third option between the traditional Humean and anti-Humean options. We can think of the Humean view as offering us a world of complete contingency in which anything could follow anything else, a view articulated in its most sophisticated modern form by David Lewis (1986). Hume himself characterised his opponent's position as one requiring necessary connections in nature (Hume 1739, p. 161) and anti-Humeans have indeed endorsed the view given to them (Harré and Madden 1975; Ellis 2001; Bird 2007). Dispositionalists have traditionally thought of necessity as the way in which causes produce their effects and such necessity is certainly anti-Humean. They posit exactly the thing that Hume claimed we couldn't see and had no idea from which to legitimately form the notion.

A middle way is possible, however. Geach (1961, p. 102) attributed to Aquinas a belief in powers having an irreducibly tendential nature. Powers would dispose towards their effects in a way that was stronger than pure contingency but less than full-strength necessity. This idea was mentioned by Harré and Madden but their book was still subtitled *A Theory of Natural Necessity*. Similarly, Bhaskar spoke of tendencies but it looks like he understood these as providing the world with necessity (Bhaskar 1975, p. 14). In short, where something has a tendency and it is in the right conditions, it *must* manifest that tendency (Bhaskar 1975, p. 214).

In recent work, we have invoked a tendential theory that is closer to the one attributed by Geach to Aquinas (Mumford and Anjum 2011a, ch. 8, and 2011b). A key difference between this account and Bhaskar's, for example, is that we assert that when a tendency is in ideal conditions for its manifestation, it still only tends to produce it. The tendential nature of dispositionality is ineliminable. Now such a tendency may be a very strong one. Tendencies come in degrees, just as a wineglass and a car windscreen are both fragile but the former to a higher degree than the latter (Fig. 8.1). Some of those tendencies, especially in ideal conditions, will almost always manifest. Almost always. But, if this is a thorough-going dispositionalism, we should not reduce the notion of tendency to that of necessity, as Bhaskar, and many other dispositionalists, seem to do. If dispositions are understood as necessities, in this sense they are merely playing the same role for which we previously invoked laws of nature. Irreducible tendencies, with a sui generis dispositional modality, allow for the possibility of a real conceptual advance: a challenge to the paradigm of what we can call modal dualism. Philosophy has had us deal only with pure contingency and necessity but neither, we assert, explain the modal nature of causation adequately.

Fig. 8.1 Vectors representing two powers with different magnitudes

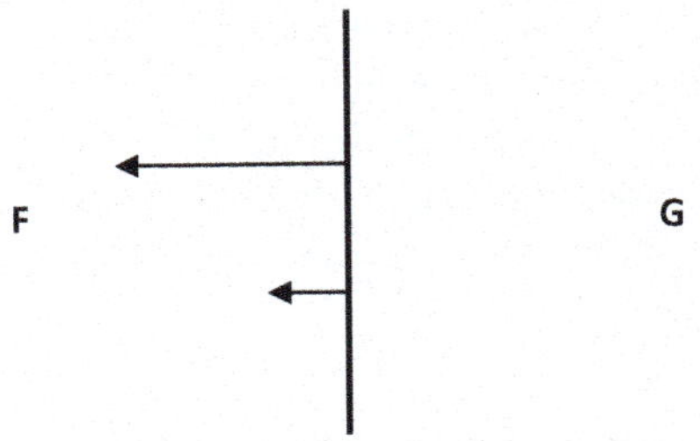

What is the argument for such a view? We argue (Mumford and Anjum 2011a, b, ch. 3) that powers produce their effects without necessitating them. The argument against necessity comes from the possibility of causal prevention: specifically, the variety we call additive interference. A set Σ of causal factors that are mutual manifestation partners for some effect E may on many occasions succeed in producing E. But on at least some occasions, Σ can be accompanied by some ϕ where ϕ is capable of preventing E even when Σ. As necessity is understood in philosophy, if A necessitates B, then whenever A is the case, B is the case irrespective of what else occurs with A. With natural causal processes, however, it looks prima facie that additional factors can indeed prevent that effect from happening. The modality of causation – what should be thought of as the natural modality – is therefore not necessity as traditionally conceived. This argument also applies to a view some call 'conditional necessity', which attempts to provide a fine-grained description of the cause, accompanied with a claim that this fine-grained cause necessitates its effect (see Marmodoro 2016; Anjum and Mumford 2018 for a historical overview of this idea). If anything, a move to a conditional necessity view should be interpreted as an acceptance of the power of our original argument, since it is an attempt to protect the cause from additional factors being added. But we have effectively offered a test of what it is to have worldly necessity: that it is robust under the addition of further factors. The conditional necessity view cannot merely assume or assert such necessity, especially if it has evaded what looks like a reasonable test of it.

The validity of this test of necessity has been challenged by Lowe (2012). His concern is that this is not in general a way to test the necessity of a conditional. We suggest that it should be possible to add anything to the antecedent and it still come out as true, if the conditional concerns necessity. But then one might add to the antecedent a condition that contradicts what else is in the antecedent, rendering such a conditional still true (trivially), and thereby passing the necessity test whether it expresses necessity or not. But what we propose is not a test of the necessity of conditionals in general but a test of the necessity of causal claims when they are posed in conditional form. And here, it seems, we needn't have Lowe's concern. Such a conditional could never be trivially true on the grounds Lowe suggests because a contradiction within the antecedent is not something that can exist in nature. Hence, there is no such antecedent that puts the necessity of the claim to the test. All such antecedents that we are testing are ones that can naturally occur and are thus contradiction-free.

We thus will retain our commitment to the necessity test for causation. This is not to say that there cannot be other tests of necessity. Perhaps some of those tests are better than ours. But, if so, let the opponents of the dispositional modality propose them and show that causation can reasonably be thought to pass that test. What we wish to warn against is the mere assumption without argument that causes necessitate their effects and, correspondingly for Humeans, the mere assumption that constant conjunction is a necessary condition for causation.

8.3 Tendential Strength and Probability

We said that a power can tend towards its manifestation to a greater or lesser degree. A power thus has a strength and this strength is something that can be gauged statistically in frequency of occurrence. We might observe, for instance, that a statistical majority of those who smoke die of a particular diseases. A study estimates that of the average 443,000 smoking-attributable deaths per year in the USA, 29% die of lung cancer, 28% of heart disease, 8% of other types of cancer, 4% of stroke and 21% from chronic obstructive lung disease (CDC report 2008). From evidence of this kind, we might reasonably conclude that smoking causes all these diseases even though there is no constant conjunction involved. The tendency towards cancer is a relatively significant one. In the cases of stroke, the tendency is not as strong in a proportional sense but given the seriousness of the effect, even this tendency of smoking can be considered a serious one. There is a stronger tendency towards lung diseases generally but even here it is far short of necessity. The tendential approach shows, therefore, that a smoker lucky enough to avoid all these illnesses in no way proves that smoking does not cause them. We cannot therefore be naïve falsificationists about causal claims.

Should we then treat these, and all other cases, as examples of probabilistic causation? Is the dispositional-tendency view one that simply renders all causation probabilistic? Such a view, which is one reason why some before us have doubted that regularity is a necessary condition for causation (Armstrong 1983, pp. 29–35), seems to fit prima facie the features of a dispositional account. Causation would indeed involve something stronger than mere contingency and weaker than necessitation. All the tendencies we get from dispositions would be probabilistic, with the strength of the probability matching the magnitude of the disposition.

However, it would be a mistake to immediately infer the ubiquity of probabilistic causation from its dispositional nature. For one thing, we do not see probabilities as always an irreducible feature of reality and there are many cases where such facts of objective chance are produced by the tendencies of the powers and their strengths. Causes may well be probability-raisers, in almost all cases, but we take it that there is a further question of why they raise the chances of their effects and our answer is that they dispose or tend towards them. This is to say that we do not want to explain tendencies in terms of probabilities because we instead explain probabilities in terms of tendencies. We think such tendencies could also give a non-trivial sense to the ceteris paribus clauses that some think can be attached to general causal claims (Mumford and Anjum 2011a, b, pp. 165–167). If ceteris paribus clauses are not indicative of the tendential nature of the claim, then we are not sure what they are meant to do instead.

If tendencies are to be the worldly grounds of the truths of probability, then they cannot be probabilistic in nature themselves. But we can see that they are not, on a true tendency account. The strength of a power is not simply the same as the probability of its effect. We measure and state probabilities on a bounded scale between 0 and 1, where there cannot be a higher probability of e than 1 or a lower probability than 0. The strength of a power, however, for at least some cases, is measured on an unbounded scale. This means that there can be more than enough to produce a certain effect. Consider the use of a brick as a paperweight, for example, which should always be more than enough power to stop the papers blowing away. And yet it is still possible to double the extent of the power by adding another brick on top of the first. While the extent of power is thus unbounded, within the theory of tendencies, it is still always the case that the classical probability of an effect is < 1, because no effect is necessitated. Hence, we can have as many safety mechanisms in an aeroplane as we like, and we can even back them all up, but we still know that the probability of safely completing the flight is very slightly less than 1. If we think of a propensity as the having of a power or tendency to some degree, then, it is clear that it is not the same as a classical probability. That is good news if you want propensities to be the ontological grounds of probability (Anjum and Mumford 2018).

Nevertheless, as well as the regular, non-probabilistic powers that tend to some degree to manifest, there are also probabilistically constrained powers that involve an objective chance of some outcome or other being realised. To understand the exact nature and usefulness of this distinction, we need to consider different ways in which statistical data can be treated as probabilistic.

8.4 Interpreting Statistical Data

It is quite natural and to a degree justified to look for causes using statistical methods. However, we think that the right methods have to be used and their limitations acknowledged. It ought also to be accepted that there are lessons to be learnt from the sort of data that are produced in most scientific trials.

For a start, we should not expect that the pure constant conjunctions, that Hume thought the starting point of causation, are going to be found in scientific tests such as randomised controlled trials (RCTs). That would require that, upon some intervention C, an effect of type E always followed. Even the most 'successful' RCTs will not deliver anything close to that (Howick 2011, Cartwright and Hardie 2012). And we should bear in mind that in the *Enquiry* Hume also invokes the idea that as well as A being followed by B, we should be able to say that without A, B would not follow (Hume 1748, section VII). Hume is sometimes taken as offering two completely opposed theories of causation here – a constant conjunction theory and a difference-making theory – and mistakenly thinking them to be the same. That would seem quite a blatant error to ascribe to Hume, given his evident power of subtle argumentation employed elsewhere.

Instead, we might think of Hume, having already acknowledged that the idea of cause is complex, as seeing that the requisite regularity took two forms. Certainly, A should be followed by B, but that also would not be enough for our idea of cause because if B was ubiquitous – and therefore followed every event – its following of A would not be adequate for us to form the idea that A caused B. What we would need, as well as every A being followed by B, would be the regularity that without A there is no B. Constant conjunction and difference-making elements of regularity make philosophical sense, taken together, therefore, as the signs of causation.

But RCTs, while testing for both elements, do not reveal what Hume is looking for in any absolute way. A typical RCT (see, for example, Broadbent 2013, p. 4 f. and Howick 2011, p. 45), we will suggest, may require division of a population into two or more groups: a treatment group that receives the intervention whose effects we wish to test, and at least one control group that receives nothing or a placebo. This type of RCT might be suitable for testing the effect of a new drug, for instance. Suppose the trial drug is hoped to aid recovery from some disease, D. Naturally, we will invoke some of the standard requirements on RCTs (Rabins 2013, p. 149). The size of the population must be large enough that we will consider any resultant data significant, and the division into the two groups must be genuinely random, so there are just as likely to be as many men in each group, as many smokers, as many optimists, as many depressives, and so on.

If the rate of recovery is higher in the treatment group than the recovery rate in the control group(s), this can be thought of as confirming Hume's second requirement, that without the cause, the effect would not exist either. The control group reveals the difference made by the drug. This is vital, since if the recovery rate of the control group were the same as the treatment group, no such effect has been demonstrated.

Say we find that 40% of people in the treatment group recover from D and this is greater than the recovery rate in the placebo group. This gives us some indication of how effective the drug is. But there is no regularity, in the constant conjunction sense, between taking the drug and recovery. Statistical regularity here means regularity over a number of instances where each instance is a full trial. Such regularities might reveal only relatively weak correlation, for instance that 1 in 1000 women taking a type of contraceptive pill develop thrombosis, and this is a two- to threefold higher risk than in women who take no such pill (Rosing et al. 2001, p. 194). Although this is a very weak correlation, it could be a regular one, found in a number of trials; and given the undesirability of thrombosis, it would then be considered a significant correlation despite its weakness. People might risk a 1 in 1000 chance of getting a headache but not the same chance of death, for example.

Even here, however, in this proportional form, we do not find the constant conjunctions quite imagined by Hume. It is unlikely that upon systematic review, we find that every trial of the same treatment reveals exactly the same recovery rate. One thing we could do is simply add together all the trial results and calculate the proportions by averaging. This might not be methodologically entirely sound, however, as not all trials will have employed exactly the same study design. Perhaps some of the trials had a treatment group and placebo group but no third control group, for instance. The systematic review would then have to be slightly less strict, therefore, and willing to overlook differences in study design, between the different trials if they are considered minor enough.

The best Humean regularities that are found in statistics, therefore, are most typically a pattern of raised incidence of E upon C, showing a significant enough 'effect', within a range of variance. Epidemiologists may be trained not to assert a causal connection between C and E even in these cases. They are scientists rather than metaphysicians, after all, and the extent of the scientific claim may stop with there being a correlation between C and E. Another pattern that could also be sought, in more sophisticated trails, is that one factor, F, correlates with another, G. This means that it could be found, for instance, that increasing the extent of F, through appropriate testing, correlates with an increased extent of G, where F and G are quantities able to vary in magnitude. And a lower extent of F could correlate with a lower extent of G. Such data is not always available, since trials might only look at a type of treatment and proportional rate of recovery. But this further information seems useful to know if, for example, we can increase the quantity of the treatment. Does increasing the dosage of a trial drug increase the recovery rate? Finding that one quantity varies with another can be important when deciding the correct dosage levels for the best treatment.

Those more philosophically minded might think it reasonable to draw an inference from such statistical evidence to the conclusion that there appears to be a causal *influence* of C upon E or of F upon G (see Lewis 2004 for one account that exploits this feature). But given the limited nature of the regularity, as we have described it, there are still some different interpretations that are possible in respect of probability assessments.

In the case mentioned, there is a recovery rate in the treatment group around 40%, in 4 out of 10. But because this is a study over a population – which is necessary for the results to be considered statistically significant – this is consistent with a number of different possibilities concerning the effects on the individuals within the treatment group. One possibility is that the drug has no effect whatsoever on 60% of the group and a strong and successful effect on the 40% who show recovery. This interpretation would fit well with the idea that causation is a matter of all or nothing. Either the effect occurs; or it doesn't, in which case no causation has happened.

Using the statistical information to guide clinical decisions, therefore, one could infer that there is a 0.4 chance of recovery with this treatment. It is important to note, however, that this probability assignment is a matter of epistemology, evidence and risk; not ontology. There is a 0.4 chance that we are right in assuming that the treatment will cure D for this patient; that is, this is how likely it is that the patient belongs to the right sub-group on which the treatment has an effect. This probability assignment is therefore about our confidence, or credence, in the predicted outcome.

An alternative is to say that causation itself is probabilistic, and that the statistics reveal that the drug raises the probability of recovering from D to 0.4. This is an ontological form of probability concerning the way some restricted part of the world is. Next, we will look at two ways to interpret statistical data probabilistically in the ontological sense: those ways being frequentism and propensity theory.

8.5 Frequentism

One key decision we have to make in interpreting statistical data is whether we should read it as frequentists or propensity theorists. On a frequentist interpretation of probability, the probability of an event is determined by its relative frequency over time. This frequency is found through a large number of repetitions under similar conditions. David Lewis (1994) spoke of naïve frequentism, which would simply be a reading off of probabilities from the distribution of events in known experiments, or from the frequency of occurrence. There are more sophisticated versions of frequentism that can be followed within a broader metaphysical framework, however (see Gillies 2000, ch. 5 and Eagle 2011, part V). If one is a four-dimensionalist, one would think that there is a frequency of distribution over events of a certain type for all times. In the entire history of the universe – past, present and future – for instance, it may be that of those things that are F, 59% of them are also G. Given this omnitemporal fact, then we should say that the chance of any F, picked at random, being G is 59%.

Needless to say, within science we do not have access to every F that is, was and will be G. Nevertheless, part of the rationale for RCTs might be that if one chooses test subjects in a properly random way, the sample of those things that are F and are chosen for the RCT ought to be more or less representative of the group of all Fs as a whole. Of course, one may get an unrepresentative sample but that would be the reason for repeating a number of studies and then reviewing them systematically. If a pattern emerges over a number of trials, then it can be thought of as indicating the approximate proportion of Fs followed by Gs.

The distinctively philosophical commitment of frequentism is that probability is constituted by the distributions of outcomes. For the naïve frequentist, this might simply be the distribution over a sequence of trials; for the more sophisticated version it would include every distribution of Gs within F. This means that probability is generated from statistics, whether known or all possible. Statistics reveals probability for the frequentist because probability is nothing more than the frequency with which the outcome occurs.

We can grant that frequencies are indeed a way in which we can come to learn about probabilities. We might see that a frequency with which a die lands 6-up is more than 1 in 6 times, for instance, over a reasonably large number of trials. This could lead us to conclude that it is a loaded die, of course, but also something about the degree to which it is loaded. Perhaps after 100 rolls, there have been 50 sixes, and then we conclude that the chance of it landing 6 is 50:50. Here, one simply records the outcomes, calculates relative frequency and draws a conclusion about probability. One has begun from a position of ignorance, but this might be just like our initial ignorance over the chance of a drug causing recovery. And for the frequentist, there is no further metaphysical fact grounding probability other than such distributions.

Frequentism faces challenges, however. Some of them concern whether facts of chance can be fixed when there is only a low number of occurrences of a phenomenon (Mumford 2004, p. 47 f.). Another problem is that the probability generated from frequency can only tell us something about a statistical average of a sample, which might not be representative for any of the individual instances from which the sample is gathered. Indeed, frequentism is too blunt to distinguish some of the different possibilities there could be when on average 40% of Fs are Gs, such as: all Fs having a 40% chance of being G, half of them having a 50% chance of being G and half having a 30% chance, or each having a separate chance within a range of 0–100%. Prima facie, it looks like the theory has no way of distinguishing these conceptually and ontologically distinct possibilities.

Now it might be alleged that the frequentist can make the requisite distinction by being adequately fine-grained about group membership. It might be said that of those things that are F, they divide into two groups, A and B, and frequency of being G in group A is 50%, while frequency of being G in group B is 30%, and again such facts of frequency are all the facts of probability. But this is not a move likely to lead to success. For a start, it seems arbitrary where to stop. Every individual could belong to ever more fine-grained sub-groups. And if one resolves the group down to the one that most fits an individual person, it is the N-of-1 group of which this individual is the only member. Probability claims are redundant in N-of-1, within the frequentist framework, given that this individual is simply either G or not. But such a terminus seems avoidable only if one takes some notion of a smallest relevant sub-grouping of F, and that seems an arbitrary matter.

If one avoids going the N-of-1 route, another objection seems to state that one individual can belong to many different groups and sub-groups at the same time. The facts of possibility do not then look fixed as an ontological fact but are, instead, merely relevant to our interests. The same individual may have a 40% chance of recovery from a disease *qua* man, but only a 20% chance of recovery *qua* male smoker, and yet a 30% chance of recovery *qua* male vegetarian smoker. Clearly, the same person can belong to a number of different sub-groups with the chance of recovery being different according to how finely grained one considers their grouping (it is the job of actuaries for life assurance companies to calculate chances based on all such factors).

8.6 Propensities

An alternative to frequentism is propensity theory. We have said that the ontological commitment of the dispositionalist is towards irreducible tendencies rather than regularities. What this will mean is that within any sub-group, no matter how small, there will be a tendency, and no more than a tendency, towards exhibiting some effect. It is important for the anti-Humean to say this. If they did not, then the Humean could easily claim that there is after all a pure, constant conjunction, of the kind they originally claimed: but it was just that the constant conjunction got masked in large, varied populations with multiple factors at work. While only 40% of the treatment group recovered in our RCT, for instance, the Humean could allege that if a person received the treatment and satisfied a number of other conditions – perhaps they are non-smoking vegetarians who exercise – then they always recover. Here is the pure, Humean regularity or constant conjunction. But this is what we deny as dispositionalists.

Now in a relatively small sample, it may be that all those who have this specific set of factors indeed recover. And the tendency towards recovery when all these factors are present may indeed be a very strong one. But perhaps if there were 10,000 people subject to all these causal influences, a handful of them would not respond in the same way. Even a strong tendency could fail to manifest itself, given enough trails. The most likely situation going back to our original recovery rate of 40%, as far as a dispositionalist is concerned, is that the varied population within the treatment group will have differing tendencies towards recovery. Some of the population will have a strong tendency to recovery, some will have little or none, and the statistic of 40% is what that whole bunch of different degrees of tendency produce taken together.

A consequence of this is that if the treatment in question becomes a marketed drug, it cannot be inferred with certainty from the outcome of the RCT exactly what effect the treatment will have on an individual considering using it. There is perhaps a 40% chance of recovery for the individual *qua* normal human being. But it is also quite likely that an individual has a number of other causal factors in operation that perhaps mean that they really have only a 20% chance of recovery. The taker of the drug has no way of knowing whether they are a statistically average kind of person. It is likely that most will deviate from the statistical average to some degree, so it is a matter only of how much so. And perhaps there is a point at which a user is so atypical that the drug has little or no effect on them, or is even harmful.

On a dispositionalist view, we understand probability to be a fact of objective chance, distinct from the issue of credence or degree of belief. And what this fact of chance resides in is the individual propensities of things: their dispositions or tending towards certain outcomes to some degree. Such probabilities can be ascribed intrinsically to individuals: such as what the chance is of this candle melting. But they can also be ascribed to whole situations, for instance when we consider the chance of a candle melting that is stood next to a furnace that is just about to be lit.

The distinction between these two types of ascription of probability is important and we should not assume that every attribution of chance is intended in the same way. There might be a probability of a woman getting breast cancer based purely on demography, ultimately based on the sort of evidence we have discussed from trial data. But perhaps given certain environmental factors to which one particular woman has been exposed, such as the presence of carcinogens, she might have a greater or lower chance of getting cancer than is typical for her demography.

However, crucial for the dispositional account is the idea of the intrinsic dispositional tendencies that an individual contributes to the overall mixture of tendencies in play. A sugar cube will have an intrinsic tendency to dissolve in liquid, for instance. Glass will have an intrinsic degree of strength. Now it can be more likely that some strong glass has been broken while some weak glass has not because whether or not any particular pane of glass breaks depends on the other factors in operation, which we understand as mutual manifestation partners. All causally active powers, operating upon one situation, are understood as mutual manifestation partners jointly producing their effect (Martin 2008, ch. 5 and Mumford and Anjum 2017). Strong glass might, for instance, break with only slight impact if the glass is subject to very low or very high temperatures. Here, the temperature can be seen as a causal contributor. The effect produced is a result of the composition of causes (Mumford and Anjum 2011a, b, chs. 2 and 4), which allows for compositional pluralism: the idea that composition need not be according to a linear function. But each particular will bring its own intrinsic propensities into that composition, even though the effect is caused jointly. The individual in our example above adds any genetic predispositions for or against cancer, for instance, plus any acquired tendencies towards it. The sugar cube has an intrinsic disposition towards dissolving but one it can exhibit only when it meets a suitable solvent.

Because such tendencies are understood as intrinsic to the individual, rather than to the individual *qua* member of a group, they are the sort of thing that might be revealed in one well-conducted experiment, in which the experimenter attempts to assemble all the right mutual manifestation partners for that disposition (Cartwright 1989, for example). If one wishes to test magnesium for flammability, for instance, one need not conduct a prolonged series of many trials. One experiment in which magnesium ignites in oxygen might be enough. And any further tests might just be to check the experiment meets the standards of repeatability, or to train budding scientists, rather than be an exercise in acquiring more data.

Statistics can be symptomatic of the intrinsic propensities of things, but they do not define or generate them. The reason for this can now be set out explicitly, though it follows from what has been said earlier. Powers compose but they do not decompose. The intrinsic tendencies of all the powers operating on a situation will compose in accordance with some function to effectively become a resultant tendency, which can then be thought of as producing the effect. The resultant power should not be thought of as some new entity. As we say, this is entirely composed by the component powers according to some function. There is no additional element other than the component powers; but in their mutual manifestation, or joint action, they can work with or against each other, sometimes in a non-linear way. This means that with any set of individual powers, each with their distinct magnitudes, there is only one resultant power that they compose. But we cannot decompose them, meaning that from any known resultant power, there is any number of component powers from which it could have been compositionally produced. We are offering this not just as an epistemic claim that from the resultant you cannot deduce the components. When powers compose they become one and initiate a process that is their mutual manifestation. In the effect, the component causes no longer have individual identities. The absence of decomposition is an ontological as well as epistemic fact.

This finding explains our problem with reading objective chance purely from frequencies. In any population, many different and varying tendencies will go into producing a distribution in a trial, for instance of Fs that are G. But we cannot decompose that distribution and ascribe it back to individual Fs, as the chance that they will be G. 40% of Fs being G is consistent with many different degrees of propensity of things that are F towards being G.

8.7 Probabilistic Tendencies

We can now return to an earlier issue: of whether all powers are effectively being treated as probabilistic in our account. We say not because a distinction can still be drawn between tendencies of a certain strength, that either manifest or don't, and genuinely probabilistic dispositions in which there is a balanced distribution of dispositions towards various outcomes. A fair coin gives a model of the latter, where the chance of landing heads or tails if tossed is 50:50, which we represent as a single double-headed vector (Fig. 8.2). Another example is a die, which disposes towards six different outcomes simultaneously – equally so if it is fair – and where the chance of all those outcomes must add up to one. As should be clear from what has been said, however, both a regular tendency and a probabilistic disposition can be understood as propensities and are an ontological basis for the assignments of objective chance.

The issue of probabilistic powers ought also to be kept separate from the issue of indeterminism. The probabilistic power of a die, which disposes in six different directions, may well be a perfectly deterministic one, if the side the die will land is fixed by all prior states. And contrariwise, the power of radioactive decay might well be entirely indeterministic, in respect of the exact moment at which the particle decays, even though it can be understood as a regular tendency, disposing towards only one outcome.

What the dispositional modality that we have advocated does allow, however, is a category of genuinely chancy or stochastic causation. When Mellor (1971) and Popper (1990) present their propensity theories, where they too understand propensities in terms of powers, they both give accounts in which chancy causation works by resolving into non-chancy causation. An F may have a 50% propensity to be a G, for instance, but what happens when it becomes a G is that it goes through a temporally extended process in which its chance of being G gets closer and closer to 1. And when that chance reaches 1, that is exactly the point at which it becomes G. These accounts resolve, therefore, into causal necessitation theories in which a cause produces its effect by eventually necessitating it. As already mentioned, this

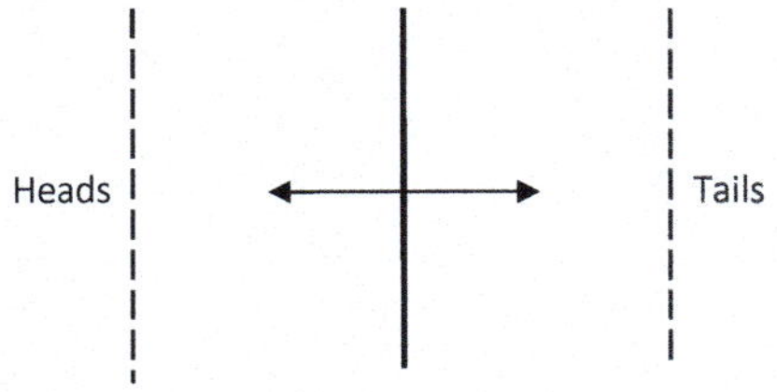

Fig. 8.2 A probabilistic power, disposing equally towards two possible outcomes. Reprinted from: Stephen Mumford and Rani L. Anjum: *Getting Causes from Powers*, Oxford 2011, p. 79, with permission from Oxford University Press

makes causation a matter of all or nothing rather than authentically probabilistic all the way through.

With the dispositional modality, however, there is conceptual space for the possibility that causes produce their effects without necessitating them. In other words, there could be a genuinely chancy case in which there was a 50% propensity towards H and H is then just caused, without any additional factor necessitating it. This goes against the engrained causal necessitarian way of thinking. But if we are to allow stochastic causation at all, it might as well be genuinely stochastic instead of just the same old necessitarian causation, thinly disguised.

8.8 A Case for Causal Singularism

We have a variety of ways of investigating the world to discover causes and probabilities. It seems plausible from what we have considered that different experimental methods suggest different thinking about the nature of causation and probability. Generalising, experimental methods to be found in chemistry and physics (see for instance Baetu 2011; McKay Illari and Williamson 2011) are conducive more to causal singularist and propensity theories of probability, for instance, while the use of RCT statistics and population data fits better with regularity theories of causation and frequentism about probability. Some call the latter covering-law models of causation. The essential difference between these two models of causation and probability is that the covering-law model takes the general causal and probabilistic facts to be primary, which are then distributed to the particular instances. The singularist takes that particular causal truths and truths of propensity to be primary, which then determine the general causal truths. We have effectively presented an argument in favour of singularism: the argument being that singular causal and probabilistic truths can reliably determine the general ones, but the general causal and probabilistic truths cannot reliably determine the single, particular ones.

Perhaps a more pressing methodological conclusion, however, is that one approach does not automatically transfer over to another (Kerry et al. 2012; Eriksen et al. 2013). When making a clinical decision about an individual treatment in medicine, for instance, it matters whether the patient has a genuine propensity of responding to it or whether there is ‘merely’ a statistical fact that others have responded. It seems that different methods give evidence of different things and that these might sometimes conflict. We can think of each experimental method as providing different symptomatic evidence of causation and propensity. But as we have shown, these different approaches may be based upon different ideas of what causation consists in and what is the nature of probabilities in nature. The theory of powers suggests reasons for a degree of scepticism about some approaches, which might be thought of as providing a rough and imperfect approximation of causation: its symptoms, at best. A singularist, propensity theory allows better for individual variation, causal complexity and context sensitivity of effects, which we take to be essential features of causation (Mumford and Anjum 2011a, b, ch. 7).

8.9 Conclusion

We have argued that regularity, often taken to be the starting point in many accounts of causation, is not something easily found in typical causal connections. Some experimental methods do not even seem to be looking for it. And those that do – population studies and randomised controlled trials – also show a more complicated picture. Such results indicate a frequency of distribution, but one that is not necessarily constant through a series of trials, and instead suggest we understand causation in terms of tendencies and probabilities.

A metaphysics of causation built upon real causal powers manages well to interpret such experimental methods. But it also fits more naturally with a singularist and tendency view of nature in which the statistical facts are generated by the individual propensities of things. This gives the dispositionalist good cause to reject frequentist interpretations of probability as inadequate both metaphysically and in terms of scientific methodology.

Acknowledgements This work was funded by the Research Council of Norway's FRIPRO scheme for independent projects.

References

Anjum, Rani L., and Stephen Mumford. 2018. Overdisposed. In *What Tends to Be: The Philosophy of Dispositional Modality*, 49–63. London/New York: Routledge.

Armstrong, David. 1983. *What is a Law of Nature?* Cambridge: Cambridge University Press.

Baetu, Tudor M. 2011. Mechanism Schemas and the Relationship Between Biological Theories. In *Causality in the Sciences*, ed. P. McKay Illari, F. Russo, and J. Williamson, 407–424. Oxford: Oxford University Press.

Bhaskar, Roy. 1975. *A Realist Theory of Science*. Leeds: Leeds Books Limited.

Bird, Alexander. 2007. *Nature's Metaphysics*. Oxford: Oxford University Press.

Broadbent, Alex. 2013. *Philosophy of Epidemiology*. London: Palgrave Macmillan.

Cartwright, Nancy. 1989. *Nature's Capacities and Their Measurement*. Oxford: Oxford University Press.

Cartwright, Nancy, and Jeremy Hardie. 2012. *Evidence-Based Policy: A Guide to Going it Better*. Oxford: Oxford University Press.

CDC – Centers for Disease Control and Prevention. 2008. Smoking-Attributable Mortality, Years of Potential Life Lost, and Productivity Losses – United States, 2000–2004. *Morbidity and Mortality Weekly Report (MMWR)* 57: 1226–1228.

Eagle, Antony, ed. 2011. *Philosophy of Probability*. London: Routledge.

Ellis, Brian. 2001. *Scientific Essentialism*. Cambridge: Cambridge University Press.

Eriksen, Thor E., Roger Kerry, Stephen Mumford, Svein A. N. Lie, and Rani L. Anjum. 2013. At the Borders of Medical Reasoning – the Aetiological and Ontological Challenges of Medically Unexplained Symptoms. *Philosophy, Ethics and Humanities in Medicine* 8: 1–11.

Geach, Peter T. 1961. Aquinas. In *Three Philosophers*, ed. G. E. M. Anscombe and P. T. Geach, 65–125. Oxford: Blackwell.

Gillies, D. 2000. *Philosophical Theories of Probability*. London: Routledge.

Harré, R., and E. H. Madden. 1975. *Causal Powers: A Theory of Natural Necessity*. Oxford: Blackwell.

Howick, Jeremy. 2011. *The Philosophy of Evidence-Based Medicine*. Oxford: Blackwell.
Hume, David. 1739. In *A Treatise of Human Nature*, ed. L. A. Selby-Bigge, 1888. Oxford: Clarendon Press.
———. 1748. In *An Enquiry Concerning Human Understanding*, ed. P. Millican, 2007. Oxford: Oxford University Press.
Kerry, Roger, Thor E. Eriksen, Stephen Mumford, Svein A. N. Lie, and Rani L. Anjum. 2012. Causation and Evidence-Based Practice: An Ontological Review. *Journal of Evaluation in Clinical Practice* 18: 1006–1012.
Lewis, David. 1986. *On the Plurality of Worlds*. Oxford: Blackwell.
———. 1994. Humean Supervenience Debugged. In his *Papers in Metaphysics and Epistemology*, 1999. Cambridge: Cambridge University Press, 224–247.
———. 2004. Causation as Influence. In *Causation and Counterfactuals*, ed. J. Collins, N. Hall, and L. Paul, 277–290. Cambridge, MA: MIT Press.
Lowe, E. Jonathan. 2012. Mumford and Anjum on Causal Necessitarianism. *Analysis* 72: 1–4.
Marmodoro, Anna. 2016. Dispositional Modality Vis-à-Vis Conditional Necessity. *Philosophical Investigations* 39: 205–214.
Martin, Charles B. 2008. *The Mind in Nature*. Oxford: Oxford University Press.
McKay Illari, Phyllis, and Jon Williamson. 2011. Mechanisms are Real and Local. In *Causality in the Sciences*, ed. P. McKay Illari, F. Russo, and J. Williamson, 818–844. Oxford: Oxford University Press.
Mellor, D. Hugh. 1971. *The Matter of Chance*. Cambridge: Cambridge University Press.
Mumford, Stephen. 2004. *Laws in Nature*. London: Routledge.
Mumford, Stephen, and Rani L. Anjum. 2011a. *Getting Causes from Powers*. Oxford: Oxford University Press.
———. 2011b. Dispositional Modality. In *Lebenswelt und Wissenschaft*, ed. C. F. Gethmann, 380–394. Hamburg: Meiner Verlag.
———. 2017. 'Mutual Manifestations and Martin's Two Triangles. In *Putting Powers to Work: Causal Powers in Contemporary Metaphysics*, ed. J. Jacobs, 77–89. Oxford: Oxford University Press.
Popper, Karl. 1990. *A World of Propensities*. Bristol: Thoemmes.
Psillos, Stathis. 2002. *Causation and Explanation*. Chesham: Acumen.
Rabins, Peter V. 2013. *The Why of Things: Causality in Science, Medicine, and Life*. New York: Columbia University Press.
Rosing, Jan, Joyce Curvers, and Guido Tans. 2001. Oral Contraceptives, Thrombosis and Haemostasis. *European Journal of Obstetrics & Gynecology and Reproductive Biology* 95: 193–119.

Chapter 9
What Powers Are Not

Elina Pechlivanidi and Stathis Psillos

Abstract This chapter analyses and criticises the idea that powers are representable as vectors. Mumford and Anjum have recently developed a vector model of powers as part of their account of dispositional causation. The purpose of this model is to represent powers as causes as well as to explain various features of their account of causation, for instance dispositionality, i.e., a sui generis type of modality introduced by their power-based ontology. In this chapter, we criticise both the claim that *powers are vectors* and the concomitant claim that the composition of causes can be understood as *vector addition*. We argue that powers cannot be thought of as even analogous to vectors, and that the vector model is simply misleading. We show that the root of the problem is in Mumford and Anjum's thought that powers have *magnitude* and *direction*.

Keywords Causation · Powers · Composition of causes · Vectors · Vectorial representation · Dispositionality

9.1 Introduction

Power is an ontological category introduced by Aristotle to explain and ground change and activity in nature. It was heavily challenged from the seventeenth century on but has regained popularity since the last quarter of the twentieth century. 'Power' is not quite defined. But its key feature, at least according to Aristotle and most of his followers, is that a power can exist unmanifested though it can be manifested under certain circumstances. Within the current neo-Aristotelian ontology, causation is widely taken to be the exercising of powers; it is the production of an effect (manifestation) from its dispositional cause(s).

E. Pechlivanidi (✉) · S. Psillos
Department of History and Philosophy of Science, University of Athens, Athens, Greece
e-mail: elinapec@phs.uoa.gr; psillos@phs.uoa.gr

A. S. Meincke (ed.), *Dispositionalism*, Synthese Library 417,
https://doi.org/10.1007/978-3-030-28722-1_9

In their recent work (2011a, 2018, Chap. 8, this volume), Stephen Mumford and Rani Lill Anjum (henceforth M&A) have aimed to articulate a power-based theory of causation, putting emphasis on the novel modal category of *dispositionality*.[1] Part of this theory is that powers are represented as *vectors* (2011a, b, Chap. 8, this volume). They maintain that the vectorial representation captures key features of powers and that it is an indispensable tool for their theory of dispositional causation. Furthermore, they claim that this model plays a heuristic role. Traditionally, powers were taken to operate with metaphysical necessity: causes were conceived as necessarily related to their effects. But M&A take powers to possess an irreducible, *sui generis*, modality: dispositionality. For them, powers dispose towards their manifestation without necessitating it; they are characterised by a *directedness* that is analogous to the *aboutness* of mental intentionality (2011a, p. 185). This view, which is central to their theory, is claimed to be brought out by the heuristic role of the *powers-as-vectors* model; this model is supposed to show how causal production does not occur via necessitation; given that causal production is not a contingent matter either, the powers-as-vectors model is taken to favour dispositionality (2011a, p. 46).

In this chapter we challenge the *very idea* of the vectorial representation of powers. We begin with the strong claim that powers are vectors and we refute it. We then show that even if we were to accept an increasingly looser association of powers with vectors, it would still be the wrong way to go. Either as constitutive of powers or as a metaphor for the composition of causes, vectors fail to represent causes even at a *qualitative* level. Finally, we claim that even the simple idea of *directedness* fails. We contend that, in the end, dispositionality does not emerge from the M&A model as, when understood in terms of directedness, it becomes poorly supported and highly controversial.

9.2 The Representation of Powers as Vectors

The motivation for choosing vectors as a tool for the representation of powers arises from pandispositionalism: the view that all properties are essentially powers, that is that they are individuated by their causal role. On this view, powers are *the drivers of change* and the grounds of (the new category of) dispositional modality, independently of the level at which they operate.[2] Hence, when a certain effect is produced,

[1]For a criticism of the modal category of dispositionality see the review of Anjum and Mumford (2018) by Ioannidis and Psillos (2019).

[2]M&A typically focus on powers and qualities that operate at a macro-level. Though they do acknowledge that the use of certain macro-properties (such as 'being cold' and 'being hot') does not agree with the way science treats the relevant properties, they take it for granted that all properties are clusters of powers and thus that at the macro-level too, insofar as objects are engaged in causal relations, there are powers that ground causal relations among them.

this occurs by virtue of the manifestation of some powers. The production of an effect depends on the way various *powers work with each other* (and sometimes against each other). Having said this, causal production does not involve necessitation. Rather, it is readily admitted that a power-based cause is not a sufficient condition for its effect (hence, it does not necessitate its effect), since it is always possible that some counteracting power may interfere in a given causal situation so that the characteristic effect does not follow (2011a, ch. 3, 2018, ch. 1; Chap. 8, this volume). This is a novelty of their view: causation is neither contingent nor necessary. As they put it:

> Causes can be thwarted, for instance, and this should show that they never guarantee their effects, even when they succeed in producing them. Causal production should not, therefore, be conflated with causal necessitation. (2011a, p. viii).

The view of causation they favour is dispositional.[3] Causes dispose towards their effects:

> Causation involves what we call a dispositional modality. Causes dispose towards their effects, where disposing towards something involves an irreducible *sui generis* modality. The modality is something between pure contingency and pure necessity and is irreducible to neither. (2011a, p. viii)

The combination of the thoughts that causation involves various interacting powers, and that a cause is something that disposes (*and no more than disposes*) towards its effect, constitute the rationale for the main representational scheme that M&A promote: that powers are vectors. Moreover, powers are thought of, albeit implicitly, as forces. Vectors are used by M&A to represent the individual acting powers that operate in a given causal situation that *push* (in a certain sense jointly) towards an effect. They are meant to capture two key features of powers, i.e. *direction* and *intensity*. As M&A put it:

> [A] power will have a direction—that towards which it is disposed—such as fragility being a disposition towards breaking. And it has intensity. A power can be more or less disposed towards an outcome, [. . .] [as in] the case of fragility, a wine glass will be more fragile than a car windscreen. (2011a, p. 24)

To thwart an initial objection, we should stress that M&A take the vector-model utterly seriously as the way to represent power-based causation. Hence it is not a merely useful metaphor. They take it to be licensed by their realist view about powers and, thus, they claim that "a realist about powers should prefer this way of representing causation" (2011a, p. 20). *Prima facie*, there is a basis for the model: powers are said to be 'directed' towards their effects and vectors have a direction; powers seem to come in degrees of strength (intensity) and vectors have magnitude.

Though we shall argue that these similarities are, ultimately, misleading and wrong, it should be noted that they are the chief motivations for M&A to introduce the idea that vectors are apt for representing powers. By representing powers as

[3]M&A use the notions of power and disposition interchangeably and, although we are in favour of a distinction between the two, it is outside the scope of this chapter to discuss this issue.

vectors, they claim that not only can they capture the idea of a power disposing *towards* a certain outcome, but also the fact that different powers can have *different dispositional intensities* regarding an outcome. In the example in the quotation above, while both objects are disposed towards breaking, the wine glass is said to instantiate the disposition of breaking in a higher intensity (it is more disposed towards breaking) than the car windscreen. According to M&A, then, were we to represent vectorially the fragility of the wine glass in comparison to the fragility of the car windscreen, the two vectors would have the same direction, as both objects are disposed towards breaking, but they would differ in their length, representing the difference in the intensity of the two powers—with the intensity of the disposition of the wine glass being represented by a longer directed arrow.

How is this idea developed by M&A? How are the vectors plotted? M&A intend to construct models of single causal situations, e.g., the cooling or the heating of a room, or the breaking of a vase. They then define the space that the vectors (the various powers of certain direction and intensity) act upon. Given that they take it that causation involves, most of the times, changes in qualities, they find Lawrence Lombard's quality space to be suitable for the role of the basis of the vector space in their model. According to Lombard, the quality space is a set S of simple static properties $\{P_0, P_1, \ldots, P_n\}$ meeting two conditions: first, it consists of mutually exclusive static properties;[4] second, "quality spaces are kinds of properties that are such that, if any object changes by losing a property belonging to a given quality space, it must come to have another property of the same kind" (1986, 113). Trying to see how this definition can be applied by M&A, let us think of a set whose members are two simple and static properties: being red (R), and being blue (B). Is {R, B} a quality space? Something cannot be both red and blue throughout and at the same time. This satisfies the first condition of Lombard's definition. Note, however, that {R, B} does not satisfy the second condition: although both R and B are properties of the same kind (being both colours), when something ceases to be red, it does not necessarily become blue. In order to conform to both conditions, the quality space S for the kind 'colour' should be expressed in the following way: S = {x: x is a colour}, where each colour is a simple and static property, and when something ceases to be one colour it must become another. In the examples they provide, M&A have not defined the quality space in this strict way, but by using pairs of contradictories, such as 'hot' (the quality of being hot) and 'cold' (the quality of being cold). Every pair of contradictories sets up a one-dimensional quality space for the model, with the two qualities representing the two extreme states that are (metaphysically) possible for a given causal situation. Comparing this to the requirement of Lombard's definition, M&A (2011a, p. 23) do acknowledge that a one-dimension quality space is actually a *spectrum* of qualities, meaning that

[4]As Lombard (1986, p. 113) says: "Static properties that objects can have and then lack [. . .] divide into kinds in accordance with the following intuitive principle: P and Q belong to the same kind just in case they are contraries and it makes good sense to suppose that a thing having P comes to have Q".

every point between the two extreme qualities represents a different quality of the same kind. However, using a pair of contradictories enables M&A to pictorially represent the quality space in a simple way, without the need to specify how fine-grained the quality space is.

To apply the model to a causal situation, one must also determine the initial state and the relevant powers that operate at that time. To fix our ideas, here is an example of a 'causal situation':

> [S]uppose you ingest calcium, which disposes towards bodily health in a number of respects, such as good bones, teeth and muscles. Let us represent this as a vector towards *F*, where *F* stands for bodily health and *G* stands for ill health. At the same time, however, you may be subject to a number of factors that dispose away from health. You could be tired, stressed, have drunk too much coffee, experienced passive smoking, and so on. (2011a, p. 27)

This is a one-dimensional quality space with two qualities, F (good health) and G (ill health), and various powers in action. In this 'causal situation', the thought is that, given all the operative powers, there is a dominant disposition towards one of the qualities of the quality space. It is assumed that there is a starting point from which causation will operate, which is represented by a vertical line drawn at the point within the quality space that captures the current state of the causal situation (e.g., the *exact* current state of one's health). On the line, the active vectors are plotted having their directions towards the quality (F or G) they dispose. Note, incidentally, that this presupposes that each operative power disposes towards one *and exactly* one quality in the quality-space. This is clearly an unjustified assumption; diet, for instance, can dispose towards both good *and* bad health. But this is an objection that we discuss later; for now, let's take the disposition towards a single quality as an idealisation of the model.

Taking a cue from John Stuart Mill's (1882) account of the *total cause*, M&A take the (total) disposing cause to be the 'sum' of the acting powers. Given that they take causes to be complex, consisting in many different powers that can work with or against each other, they claim that the way to understand this 'sum' of powers is via *vector addition*.

The use of vector addition underlines the contrast between pandispositionalism and the traditional view about dispositional causation. Traditionally powers are taken to require a stimulus and some background conditions to manifest themselves. But on pandispositionalism, stimuli must themselves be powerful and this is, at least *prima facie*, in favour of the M&A project: being realist about powers, they do not hold a distinction between the single power that becomes manifested and the rest of the conditions that enabled it to do so. As they put it:

> Metaphysically, we should judge them [the powers] on a par to the extent that they all contribute. A distinction between causes and background conditions cannot have any real ontological strength because the effect is not triggered until they are all present, which can in any case be a momentary matter. [. . .] The distinction between causes and background conditions is not an ontologically grounded one, but rather a pragmatic or epistemological one. (2011a, p. 32f.)

Thus, instead of passive powers being stimulated, C. B. Martin's (2008) idea of mutual manifestation partners looks more conducive to the M&A view.[5] Under this understanding, the total cause consists in multiple powers *working* together. M&A argue that the model has the advantage of presenting the composition of the causes for each individual causal situation: the cause is captured by the resultant vector **R**, as the outcome of *vector addition*.

Vector addition is taken as the *tour de force* of the M&A model. According to them, the vectorial character of powers reveals that they convey only a *tendency* towards an effect; the vector of the resultant power **R** captures the dispositional modality of power-based causation. More specifically, the point within the quality space that coincides with the end-point of the resultant vector represents the disposition that the certain causal situation has the tendency to manifest. It does not represent what *will* or must happen given the specific component powers; it only represents what is disposed to happen (2011a, p. 175).

On the M&A view, then, the component powers compose into the (resultant) dispositional cause. However, they take it that component powers are as real as the resultant power. They claim that the situation is analogous to the statue and the clay case, so both the component *and* the resultant powers exist as not entirely distinct existences, nor in a part-whole way; rather they stand in a *composition* relation. Note, however, that the component vectors represent the powers that act in the specific causal situation; while, the resultant power is an unmanifested power, since the resultant vector represents nothing more than a tendency (disposition) towards an effect (2001a, p. 175). This way to view things creates a conundrum vis-à-vis the relation between the powers. If the component powers are actually operating, i.e., if they are doing their work, and if the resultant power is composed of the component powers, how is it possible for the resultant power to be a mere tendency, that is, to be non-acting? What more would be needed for the resultant power to become manifested than the manifestation of the component powers? To avoid this conundrum, we should either assume that the resultant power is an additional (emergent) power which requires the presence of another manifestation partner for it to become manifested; or we should assume that the activation of powers does not consist in their manifestation. If we follow the first horn, we are in need of an account of the supposed additional power that is needed, since it seems clear that all the powers necessary for the effect are already present. If the fragility and the striking of a vase (being the component powers) are manifested, then what is the new resultant power? And why isn't it manifested?[6] If we follow the second

[5]Martin says: "I have been talking as if a disposition exists unmanifested until a set of background conditions is met, resulting in manifestation. This picture is misleading [. . .] A more accurate view is one of a huge group of disposition entities or properties which, when they come together, *mutually manifest* the property in question; talk of background condition ceases, replaced by talk of power nets" (2008, p. 50).

[6]Note that the manifestation of these two powers ('fragility' and 'striking of a vase') is the breaking of the vase, which (even if we assume it to be a power) is manifested. That the fragments of the broken vase might have a new (unmanifested) power, e.g., to cause injury if they are stepped on by

horn, it seems we are committed to the implausible claim that although a power is operative or activated, it can still exist unmanifested.

Let's leave all this to the one side and focus on the main idea, viz., that powers can be represented vectorially. In the sequel, we pose two questions concerning the vectorial representation of powers, both of which are answered negatively:

- Are powers vectors?
- Can they be usefully viewed as *in some sense* analogous to vectors?

9.3 Powers Are Not Vectors

In his *Process and Reality*, A. N. Whitehead put forward the claim that *all things are vectors* (1929, p. 309). At the final stage of the presentation of the vector model, M&A reflected on this claim and noted:

> What exactly Whitehead meant by this, we may never know. His work defies simple and unequivocal interpretation. But we have offered a model for causation in which Whitehead's statement would make some sense. All things have properties and all properties are powers, for the pandispositionalist. If such powers are understood as directed towards a manifestation, with a degree or intensity, then all things become vectors or at least they can be represented as such. (2011a, p. 45f.)

This may well justify us in attributing to them the strong claim that *powers are vectors*.[7]

9.3.1 Power-Vectors Are Not *Vectors*

Geometrically, a vector can be represented as a directed line segment with an arrow indicating the direction and whose length is the magnitude of the vector. In physics, vectors have been used for representing forces. Forces have magnitude and direction, and they add vectorially. In the M&A model vectors are used to represent the exercising powers that operate upon the quality space (2011a, p. 24). But are powers *vectors*? Do they add vectorially?

---

naked feet, may well be true but it is hardly a resultant power of the foregoing components unless we also assume as a further component the (passive) power of human flesh to be injured by broken glass. But given the activation of the new 'component' power, the new resultant power (to cause injury) would be manifested. Many thanks to Anne Sophie Meincke for pressing this point.

[7]In their (2011b), M&A seem to stress more the heuristic value of the vector model for representing powers-based causation. Even there however they take it that the representation of powers as vectors 'shapes' the way we think about powers. They cite Hitchcock (2007, p. 69): "the way in which we choose to represent some phenomenon can shape the way in which we think about that phenomenon" (2011b, p. 55).

For something to be a vector it is not enough that it has a direction and a magnitude. A moving car has both but it is not a vector. Mathematically, vectors are defined in vector spaces. Let V be a set on which two operations are defined: (1) *vector addition*, which combines two elements of V and is denoted by "+", and (2) *scalar multiplication*, which combines a complex number with an element of V and is denoted by juxtaposition. V, together with the two operations, is a *vector space* over the set of complex numbers $\mathbb{C}$ if the following properties hold:[8]

Additive Closure of Vector Addition
If $\boldsymbol{u}, \boldsymbol{v} \in V$, then $\boldsymbol{u} + \boldsymbol{v} \in V$.

Scalar Closure of Vector Addition
If $\alpha \in \mathbb{C}$ and $\boldsymbol{u} \in V$, then $\alpha\boldsymbol{u} \in V$.

Commutativity of Vector Addition
If $\boldsymbol{u}, \boldsymbol{v} \in V$, then $\boldsymbol{u} + \boldsymbol{v} = \boldsymbol{v} + \boldsymbol{u}$.

Associativity of Vector Addition
If $\boldsymbol{u}, \boldsymbol{v}, \boldsymbol{w} \in V$, then $\boldsymbol{u} + (\boldsymbol{v} + \boldsymbol{w}) = (\boldsymbol{u} + \boldsymbol{v}) + \boldsymbol{w}$.

Additive Identity
There is a zero vector, $\mathbf{0}$, such that $\boldsymbol{u} + \mathbf{0} = \boldsymbol{u}$, for all $\boldsymbol{u} \in V$.

Additive Inverses
If $\boldsymbol{u} \in V$, then there exists a vector, $-\boldsymbol{u} \in V$, so that $\boldsymbol{u} + (-\boldsymbol{u}) = \mathbf{0}$.

Distributivity across Vector Addition
If $\alpha \in \mathbb{C}$ and $\boldsymbol{u}, \boldsymbol{v} \in V$, then $\alpha(\boldsymbol{u} + \boldsymbol{v}) = \alpha\boldsymbol{u} + \alpha\boldsymbol{v}$.

Scalar Multiplication Associativity
If $\alpha, \beta \in \mathbb{C}$ and $\boldsymbol{u} \in V$, then $\alpha(\beta\boldsymbol{u}) = (\alpha\beta)\boldsymbol{u}$.

Identity Element of Scalar Multiplication
If $\boldsymbol{u} \in V$, then $1\boldsymbol{u} = \boldsymbol{u}$.

Distributivity across Scalar Addition
If $\alpha, \beta \in \mathbb{C}$ and $\boldsymbol{u} \in V$, then $(\alpha + \beta)\boldsymbol{u} = \alpha\boldsymbol{u} + \beta\boldsymbol{u}$.

Something, then, is a vector by being a member of a vector space V.

The properties expressed above implicitly define *vector addition* and *scalar multiplication*.

The *norm* or *length* of a vector is an important type of function that can be defined on a vector space. It is a function satisfying the following:

1. $\|\boldsymbol{u}\| > 0$, when $\boldsymbol{u} \neq \mathbf{0}$ and $\|\boldsymbol{u}\| = 0$ iff $\boldsymbol{u} = \mathbf{0}$.
2. $\|\alpha\boldsymbol{u}\| = \|\alpha\|\|\boldsymbol{u}\|$, for any scalar a
3. $\|\boldsymbol{u} + \boldsymbol{v}\| \leq \|\boldsymbol{u}\| + \|\boldsymbol{v}\|$

[8]Bold letters denote vectors and Greek italic letters denote scalars.

A basis for a vector space is a linearly independent set of vectors, such that any vector in the space can be written as a linear combination of the elements of this set. In most problems in physics and mathematics, the choice of an appropriate coordinate system provides great computational advantages. In the case of the usual two- and three-dimensional vectors, it is typical to represent an arbitrary vector as a sum of unit vectors, where the unit vector has the same direction as the given (nonzero) vector $\boldsymbol{u}$.

A two-dimensional vector $\boldsymbol{u}$ whose initial point is at the origin (0, 0), can be uniquely represented by the coordinates of its terminal point (u_1, u_2). This is the component form of a vector **u**, expressed as $\boldsymbol{u} = \ < u_1, u_2 >$, where u_1 and u_2 are the components of **u**. The component form of the vector with initial co-ordinates $P = (p_1, p_2)$ and terminal co-ordinates $Q = (q_1, q_2)$ is $\boldsymbol{u} = <q_1 - p_1, q_2 - p_2> = <u_1, u_2>$. Finally, the magnitude of $\boldsymbol{u}$ is given by the following expression:

$$\|\boldsymbol{u}\| = \sqrt{(q_1 - p_1)^2 + (q_2 - p_2)^2} = \sqrt{u_1{}^2 + u_2{}^2}.$$

Given such a definition of vectors and vector spaces, it starts becoming obvious that powers *qua* vectors (from now on: *power-vectors*) do not fit this description. Quality spaces, where power-vectors live, are not vector spaces. And since vectors can exist only in vector spaces, powers are not vectors.

But could it be argued that there is a sense in which power-vectors exist in a vector space? After presenting the idea of powers having vectorial features, M&A claim that:

> At this point, we are thus able to make use of another notion that comes from vectors as used in physics, namely, vector addition. Powers can combine with additive, and sometimes subtractive, effects [...]. When they do so, they may overall dispose in one direction or the other. [...] Vector addition, in any case, can proceed in a rough-and-ready way without having to add and subtract the numbers [these that correspond to the intensity of the power-vectors]. We can perform a simple analogical addition by placing the tail of one vector on the head of another. (2011a, p. 28)

To perform a vector addition as described here, it is required that powers satisfy the associativity and the commutativity of vector addition. Is adding power **a** to power **b** the same as adding **b** to **a**? Or is it the same for power **a** to be added to powers **b** and **c**, as with power **c** to be added to powers **a** and **b**? Powers do not consistently obey the properties of vector addition. It's easy to see that the order by means of which two or more powers combine may lead to different manifestations.[9]

[9]Here is a trivial example: the power of drying clothes and of washing clothes. Exercising first the power to wash clothes, and then the power to dry them has markedly different results from exercising first the power to dry clothes and then the power to wash them. More interestingly, suppose that a swimmer swims 50 metres by doing the first 25 metres breaststroke and the second 25 metres freestyle. And then again that she does the same 50 metres in the reverse order. In general, the energy expended will be different in the two trials (the means or path which the swimmer takes to complete the 50 m will affect the amount of energy they use to complete the trial).

M&A are silent on this matter. But let us assume that power-vector addition does satisfy the two foregoing laws. How much of the vector theory does the power-vector model take? It's by no means clear. To be sure, the M&A model also includes the zero vector. As they put it:

> There is a special case of vector addition that is very important to distinguish. This is a case where the dispositions towards *F* and the dispositions towards *G* balance out perfectly, such that the resultant vector is neither directed towards *F* nor towards *G*. Where we have perfectly counterbalancing dispositions in this way, we say that we have a zero resultant vector. (2011a, p. 29).

But no further information is given as to which of the listed properties of vector spaces are satisfied by the power-vector addition. In the absence of this information, it is dubious, to say the least, to claim that power-vector addition is proper vector-addition. In any case, it should be clear by now that if the claim that power-vectors are proper vectors is to be taken as at least prima facie plausible, power-vectors should be characterised by more than 'direction' and 'magnitude'. As we have seen, they are not.

To a large extend M&A focus their attention on one-dimensional quality spaces, which makes vector addition easier. However, this already creates a problem since it requires that *every* kind of quality creates a single dimension. Given a model of a one-dimensional quality space, the power-vectors that operate in it can be seen as being necessarily one-dimensional (see Fig. 9.1). One-dimensional power-vectors are not 'free vectors'; they are supposed to be represented by localised line segments, the origin of which is a point in the quality space. For instance, M&A say:

> If we are trying to represent a causal situation with respect to, for instance, the temperature of someone's hand, then the vertical line represents the temperature of the hand at the starting point such that it could become warmer, if the situation moved towards *F*, or colder if the situation moved towards *G*. (2011a, 24)[10]

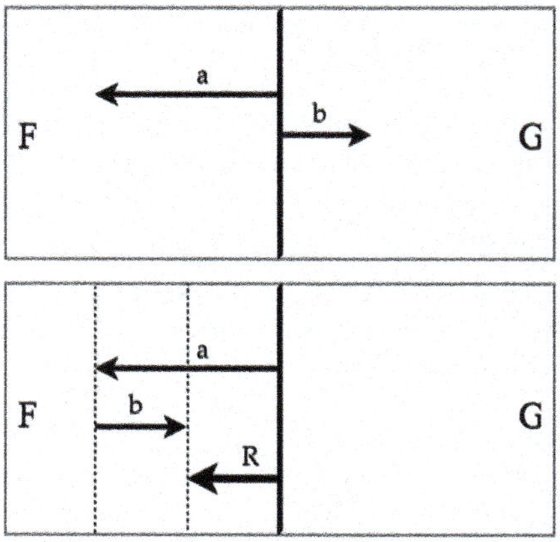

Fig. 9.1 A one-dimensional vector model of two powers at work and their addition, according to Mumford and Anjum

[10]Let's forget, for the time being, that temperature is scalar; or that energy—which actually gets transferred between the hand and the environment—is a scalar quantity. Let's play along and assume a heating power (whatever that is!) which is vectorial (*qua* power).

Here is a crucial question: what are the identity-conditions of power-vectors? When it comes to vectors, things are straightforward. Two vectors are equal if and only if they have the same direction and magnitude. Besides, any vector can be translated to the single vector that has the same direction and magnitude and its starting point is located at the origin of the coordinate system. Take velocities (or any other vectorial quantities). We can certainly tell when two velocities are the same by saying that they have the same direction and magnitude (that is, they are the same vectors—*in a given vector space*). But that's exactly what we cannot do with powers: we cannot simply tell that two powers are the same iff they have the same direction and magnitude. Hence, the whole aim of using power-vectors qua vectors to identify powers collapses.

To see this, let us suppose that we stipulate that the origin of the 'coordinate system' of the model is the point of the initial state (whichever that is) of the causal situation. Assume that a power-vector **a** (e.g., the 'power-vector' representing a power to heat the hand) has the same direction *and* magnitude with another power-vector **a**′. Plausibly, **a**′ might correspond to a different power, e.g., the power of the hand to be warmed up. If **a** and **a**′ were vectors they would have to be the same vector; but *qua* power-vectors, they are different since they represent different powers.[11]

Conversely, it's plausible to think that the same power-vector can be represented by different vectors. This is because there is no fact of the matter as to what the 'direction' of the power is and how it should be plotted in the quality space. One power-vector **a** might be plotted with direction α; but the very same power-vector might have been plotted with direction α'. This is one and the same power (hence, power-vector); but it is represented by different vectors, since they have different directions. Should the objection be that powers directed towards the same quality must have one and the same direction, the reply would be that this is not obviously so. It is possible for one and the same power to be represented by different power-vectors of the opposite direction (and even of different magnitude).[12]

Hence, distinct power-vectors can represent the same power. And conversely, distinct powers can be represented by the same power-vector. So if powers are power-vectors, we cannot tell, in principle, when two powers are the same or not.

---

[11]Could it be that these two powers (the power of something to warm something else up and something's power to be warmed up) are directed to the same quality, viz., the warmth of the hand, or even the increase of temperature of the hand? Even if we were to grant this, it would still be the case that these are distinct powers since for the hand to end up being warm both would be needed.

[12]An example we discuss later regards fragility. One can refer to fragility as 'the power of something to break easily when it is knocked', or as 'the power of having low resistance to a knock'. Both descriptions refer to the same quality of a fragile thing; however, were this quality to be represented as a power-vector, this can be done on the basis of either of these descriptions, i.e. as a power-vector of large magnitude representing the 'easily-breaking' behaviour, or as a power-vector of small magnitude representing the 'low-resisting' behaviour. These two power-vectors should have different directions. In the case of the hand discussed above, the power of hand to be warmed can be represented as being directed to an increase of the temperature of the hand or to a decrease of temperature of the surroundings.

More can be said against the power-vector model. For instance, there is no stipulation of a coordinate system. Could it be argued that such a system is implied, as it were, on the basis of a given initial point within the quality space, and the postulation of different dimensions as corresponding to different quality kinds? Even if this suggestion could be defended (as we are about to see, there are no unit vectors; nor a basis), it could only work with single-dimensional spaces and power-vectors. Yet, apart from the detailed cases of power-vectors that operate within a one-dimensional model, M&A talk about circumstances that more than one kind of quality can be affected by a collection of acting powers, suggesting that their model can be expanded to having more than one quality-dimensions. This is an example (originated in Geach 1961, p. 102) that M&A presented as a way to prompt the idea that powers interact with each other in multiple dimensions:

> [A] room contains both a heater and a cooling air conditioner. We will make the case two-dimensional. The heater can, let us say, warm the room to 25°C within an hour with dry air. The air conditioner can cool the room temperature to 10°C within an hour with slightly damp air. Suppose both the heater and the cooler are left on and the powers along the hot-cold and dry-damp dimensions are inseparable. We can still, nevertheless, understand such complex powers along the lines of vectors, and understand their combined effect along the lines of vector addition. (2011a, p. 44)

This example motivates a version of the model that consists in positioning two pairs of extremes in what seems to be a perpendicular position relative to each other, while the initial point is now a single dot on the two-dimensional vector space, on which the power-vectors of the 'power to warm' and of the 'power to cool' are plotted on a certain angle to each other (Fig. 9.2).

This set up generates many questions about the way it is constructed. For instance, how are the exact positions, directions, and magnitudes of the power-vectors to be specified? But, assuming it for now, it is hard to see how *any* system of co-ordinates can be associated with it. And given that a model such as this, if it is to be realistic at all, should allow for a number of dimensions (pairs of extreme qualities), it makes no sense whatsoever to have a formula for 'calculating' the resultant power-vector. It makes no sense to ask what the component form of the vector or its norm are. In fact, we think there is no fact of the matter. Because, for all we know there is no reason to think that there is a *correct* way to describe the operation of two powers as being additive. Why, for instance, not think of the interaction between powers as being represented by the *external product* of the vectors?

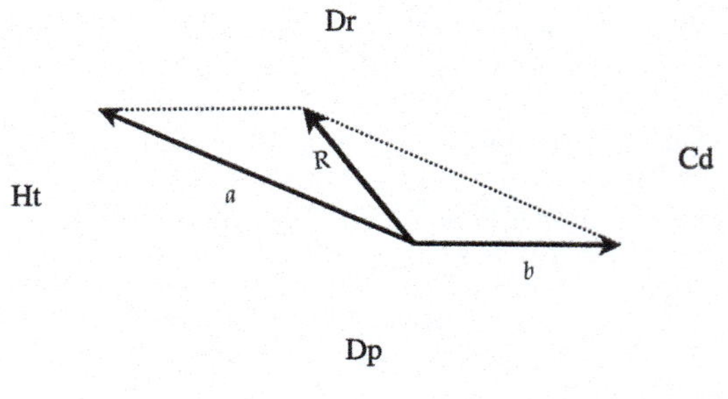

Fig. 9.2 Vectors within a two-dimensional quality space, according to Mumford and Anjum (see Fig. 2.13 in Mumford and Anjum 2011a, p. 45)

Without a well-defined coordinate system, it is impossible to locate any power-vector in a multi-dimensional quality space, because no exact angle of a power-vector can be determined. Furthermore, while any vector that forms an angle relative to the axes of its coordinate system can be analysed into component vectors that are parallel to the basis vectors, such a notion is non-existent in the power-vector model. Within a coordinate system, the basis vectors have another role too: they provide the unit upon which a scalar multiplication of the vectors parallel to them can operate. Such a unit is absent from the power-vector model. But even if such a unit were provided, there is another matter to be reckoned with: taking into account that the different dimensions of the M&A model represent kinds of qualities (e.g. temperature, humidity etc.) that are measured in different 'units', how meaningful would it be for a power-vector to be located in the 'unification' of two, or more, incompatible quality-dimensions? What sense does it make for vectors operating in different dimensions to be added?

Hence, a multi-dimensional model for power-vectors is impossible on the basis of (i) the impossibility of a well-defined coordinate system, (ii) the absence of basis vectors and measuring units, and (iii) the incompatibility of the various dimensions regarding measuring units of the qualities involved.

It should, therefore, be obvious that the power-vector model for causal powers cannot be a vector space. Powers are not vectors. To sum up, we don't know how to understand power-vector addition; we have no clue as to whether any of the properties of vector spaces are satisfied; we have no way to define the norm of vectors and no way to specify a unit vector. These are all technical limitations that the power-vector model faces. But the technical limitations are the consequence of a fundamental mismatch: powers are not cut out to be vectors after all. What remains, then, is to look into the weaker claim that such a representation of powers via power-vectors can be grounded on a (weak) *analogy* between powers and vectors.

9.3.2 Powers Are Not **Analogous** *to Vectors*

It might be that the only elements of vectors that are needed for the M&A model to work—for purposes of philosophical illustration—are the magnitude and the direction (plus the addition law, in *some* sense). But this weaker view cannot be supported either, as it leads to problematic metaphysical implications.

The thought that powers have an intensity is something that a power-theorist could accept. After all a substance can be more or less poisonous, more or less soluble, more or less flammable, or more or less fragile than another. This indicates that powers admit of degrees: x might be more, or less powerful than y to F. Therefore, one might think, something can have a power to F to a certain intensity or strength.

Indeed, this seems to be the case with 'garden-variety' macro-level powers, which are the main examples that M&A use to ground their theory. They do acknowledge that powers at this macro-level are collections of more fundamental (micro-)powers. However, their view takes it that if causation operates at all levels, so do powers. This focus on macro-level causally potent powers commits the power-theorist to properties that are to be found in folk causal explanations, raising issues about the credibility that a power-based theory may have when it comes to physics. Referring, for instance, to *the power to warm* and *the power to cool* would clearly not be accepted as being scientifically accurate. The 'cooling' power and the 'warming' power are processes of energy transfer among objects with different temperatures. And most of the quantities that M&A think as vectorially representable powers are *scalar*. But let's not dwell more on it and just examine the extent to which, if at all, garden-variety powers have 'magnitude' and 'direction'.

In those cases, where we have an intuition that powers are more or less intense, how are we to determine the intensity of a power and represent it by means of length? Accuracy in ascribing the magnitude of each power-vector is important for vector addition. Any difference in the length of the power-vector entails a difference regarding the ending point of the power-vector within the quality space; and hence a difference in intensity. But this, in turn, indicates a different possible state that the causal situation can acquire once the manifestation has occurred and, ultimately, a different resultant disposition. The requirement of accuracy is, therefore, very crucial for the purposes of the model, if it is to offer anything other than vague, arbitrary and purely qualitative representations of the causal situation. But the very idea of a magnitude requires that powers are quantifiable; and more specifically, it requires that there is a unit-power. For only the ascription of a unit can result in the representation of a power in terms of magnitude.

Not all powers are quantifiable. Therefore, there is no way to determine the length of the directed segments in the model. While, for instance, we can claim that given the same amount of pressure a glass object has a greater tendency to break than a similar ceramic object, there is no certain (or absolute) intensity associated with the manifestation of fragility. Something is simply characterised as fragile because it is *more disposed to break* in *comparison* to other things, or it is *disposed to break to different degrees* depending on the amount of pressure it will receive. Hence, a power like the fragility of an object is context sensitive, and doubly so. For one thing, it is a comparative power, therefore it needs a contrast-class. For another, it can be manifested in different degrees depending on the force that is exerted on the surface of the object, even if the intrinsic properties of the object are unchanged. The vector model is incapable of capturing this context-sensitivity via the intensity that is ascribed to powers. For instance, there is no way to show that the tendency of a glass to break (consequently, the magnitude of the corresponding power-vector) varies depending on the magnitude of the pressure that is exerted on it. As it has been acknowledged already by D. Manley and R. Wasserman (2007, p. 70), as long as there are dispositional predicates displaying context sensitivity, degrees, and comparative use of powers, a scale for each power is required, "along which objects can be compared, and by reference to which the context-dependence of dispositional predicates can be explained".

The demand for a unit of measuring the intensity of a power might be taken to be too strong by M&A, as they do not require of their model to provide any quantitative results, but to *qualitatively* represent what is disposed to happen in a causal situation. For these purposes, it might seem, depicting *roughly* the intensity of a power would be adequate. After all, they take it that although the power-vector model does not introduce any units for measuring the intensity of a power, it still succeeds in representing it. It is, however, just a lucky accident that the M&A model seems to work for some cases. This happens because the powers being involved in the chosen examples are associated with two kinds of qualities: either they are used to capture a change in some scalar quantity (e.g. a change in the temperature of the room by 10 °C), or to represent something that is already a vectorial quantity (e.g. the force that a book exerts on the surface of the table). These are exceptional cases. However, the majority of powers cannot be so represented. Most powers are not even intuitively comparable. Which arrow should be longer in the model for the combustion of a match? Is the power of the flammable material smaller or larger in intensity than the power of the oxygen? There is simply no fact of the matter. And how are arrows to be plotted in the case of the repulsion between two charges? Or the attraction between two masses?

Perhaps all that M&A care about is plotting the resultant disposition. But then it is *totally* irrelevant how we plot the 'intensities' of the various component powers, provided we get right the prevailing intuition about what the dominant disposition is. Furthermore, the power-vector model heavily overdetermines the causal situation. Suppose that in a causal situation, all the powers that are relevant to the breaking of a vase are acting. If there is a dominant disposition towards the manifestation of breaking, the only constraint on picking the 'intensities' of the component powers is that when they are 'added', the 'resultant' is pointing towards the breaking side of the quality-space. The very idea that powers have specific 'intensities', *whatever that means*, is redundant and irrelevant. In the end, it is totally clear that the analysis of the dominant disposition of a causal situation to component powers with certain 'magnitudes' is completely arbitrary.

But leaving aside the idea of the 'intensity' of a power, can we at least make good sense of the *direction* of powers? The application of the notion of direction to power-vectors is very problematic. It is enough to show that it creates inconsistencies for which any attempt to be resolved is purely *ad hoc*. The core of the problem is in the thought that even if we were to grant that all powers are directed towards their manifestation, this 'directedness' can be seen as analogous to the *direction* of a vector.

One way to see this is by examining the way that the M&A model represents non-acting, that is unmanifested, powers. For they too have a tendency towards their manifestation and therefore they should be represented as power-vectors in the relevant quality space (even though they are not acting). To fix our ideas, let us think of the toy example of an unburned inflammable match. Since flammability is a power, it should be represented as a power-vector within the relevant quality space of the combustion of the specific match. This quality space is one-dimensional and has two boundary qualities: *not burned* (~B) and (say) *totally burned* (B_t). The initial point of this causal situation is on ~B, given that the match is unburned. Suppose that in the environment of the match there is no oxygen. Then the only power-vector to be plotted is the flammability-vector directed towards B_t. Even in this isolated situation, following the norms of the model, what is being represented is the dominant tendency of the current situation; that is, the tendency of the match to become burned just by plotting its (power of) flammability. This example is analogous to other dubious powers that are supposed to be directed towards their manifestations, such as that of every living thing being disposed towards its death, or the sunrise being disposed towards the sunset. The possibility of unmanifested powers suggests that an understanding of tendency or dispositionality in terms of directedness is problematic.

M&A themselves have noted the inability of the power-vector model to represent consistently unmanifested or single powers. By emphasizing that powers require manifestation partners in order to manifest themselves, their suggested answer is based on the following idea: as long as there is some other power, or powers, whose existence is required for the manifestation of power *p* to occur, if a power *p* remains unmanifested, this must be because there is an equilibrium that keeps it unmanifested. The motivation of this idea is by the following example:

> Dynamite [. . .] has an explosive disposition which seems to manifest only when it is ignited. Without stimulation, it might be thought, it does nothing. On further inspection, however, we find that dynamite's explosive power comes from it being made up in three parts of nitroglycerin, a substance that is so explosive that its disposition has to be counterbalanced by one part diatomaceous earth, which gives it a countervailing power of stability. [. . .] [So], nothing happens in respect of the dynamite's power of explosiveness only because it is counteracted by other hidden powers. When the fuse is lit on a stick of dynamite, this is actually the addition of a further power, which then takes the situation out of the equilibrium (2011a, p. 36f.).

For them, cases like the unmanifested power of explosiveness are being misrepresented. These "unknown, hidden, or taken-for-granted", powers have to be represented in the model as power-vectors that have opposite direction and equal magnitude to the unmanifested power (2011a, p. 36). As a result, the powers cancel each other out, their resultant disposition of the causal situation being of zero intensity, thereby allowing—according to M&A—the causal situation to remain unaltered, as if nothing happens. Although they present this as a special case, there is no reason to think that this description does not generalise to *all* possessed but unmanifested powers. After all, according to pandispositionalism, the components of (resultant) powers are powers too. Thus, following the picture that is sketched here, when a power is unmanifested, the component powers are in an equilibrium state. Suppose the fragility (f) of a glass is unmanifested, viz., the glass, though fragile, is not broken (~B). The way that M&A describe the situation of the unmanifested power of fragility is the following:

> [F]ragility on its own seems to do nothing, but isn't this only because the fragile object also possesses some countervailing stability and elasticity, which holds that object together enough until some knock is sustained? (2011a, p. 37).

Is this plausible at all? On the one hand, it inflates unnecessarily the metaphysics of powers, as it leads to the supposition of additional powers to explain (by counteracting to) unmanifested powers. On the other hand, it is inconsistent with the setup of the very M&A model. Figure 9.3 illustrates the application of the model for the example of unmanifested fragility:

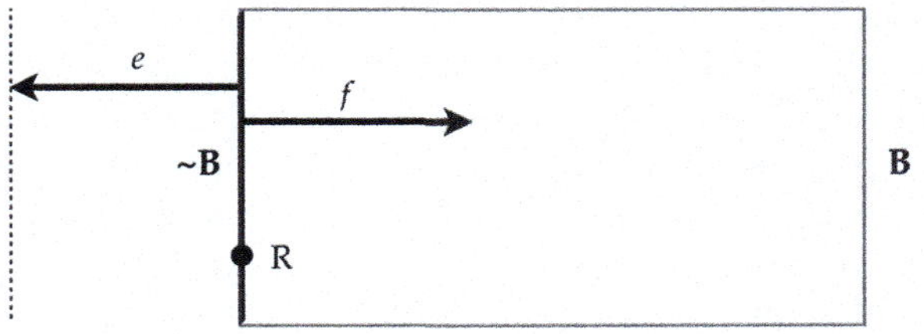

Fig. 9.3 The equilibrium of unmanifested powers, according to Mumford and Anjum

The power-vector **e** represents the power of elasticity mentioned in the quotation above, as the one that countervails **f**. The view that powers have directedness even when they are not manifested leads to postulating this countervailing power. However, the restriction in what the direction of **e** should be, so as to explain the equilibrium situation, immediately creates a tension between this setup and the notion of directedness. For, what is **e** directed towards? Towards more elasticity? Or towards more stability? Does it 'push' the glass towards being as much elastic as it is fragile? Then why is the glass characterised as fragile? An object that is characterised as very fragile can be said to be minimally elastic. Then by 'fragility' and 'elasticity' we seem to denote one and the same power, and it is for purely pragmatic reasons that we opt for using one name over the other.[13] Not only does the M&A solution to the alleged directedness of unmanifested powers not solve the problem with the notion of directedness that it assumes, on the contrary; it actually accentuates it. After all, positing a counteracting power-vector **e** is clearly incompatible with the setup of the power-vector model, as **e** appears to be operating outside the quality space {broken, unbroken} where there is no quality towards which to be directed.

Moreover, if we consider that in every application of the model the resultant power R does not differ ontologically from any other power, in a case of equilibrium where the resultant R has zero 'intensity', (it is represented by the 'zero' power-vector), this power does not have any direction either. Since R does not have either intensity or direction, it does not differ (and hence cannot be distinguished) from any other zero-resultant power.

When causality occurs, it does by means of what is described as 'the resultant power'. Thus, given the epistemic inaccessibility to the component powers, one can only intuitively determine the direction of each component power. Maybe in some cases, 'fixing' the direction of the contributing powers is more straightforward than others. However, in examples where the directedness is context-dependent, one power can have different directions: while an amount of serum cobalamin (vitamin B12) has the power to treat vitamin B12-deficiency, disposing towards better health,

[13]One could point out that there are certain different properties to be identified in a glass's material structure, some of which support fragility ('easily-breaking behaviour') and some of which support elasticity ('resisting-to-break behaviour'). But it should be noted that the qualities of a (homogenously made) glass are not divided to those that support fragility and those that support elasticity. All there is in the glass as a support for its dispositions is its lattice-like molecular structure. In fact, the linear elasticity of the glass is a key factor determining fragility.

in different circumstances the same amount of vitamin B12 has the power to raise the serum cobalamin levels so high that there are pathological consequences—thereby disposing towards deteriorating health. This example shows that whereas a component power has some directedness towards a certain manifestation, in this case *the power towards raising the levels of serum cobalamin*, in different contexts the very same power has to be represented as having different direction. Consequently, it is wrong to associate the dispositional character of powers with the view that powers have any particular *direction*.

The root of the problem, it seems to us, is that the sense in which the power is 'directed to its effect' is, at best, metaphorical. A vector is *directed* quantity; this means it has a direction. It does *not* mean it is directed towards anything. To find the direction of a vector quantity, e.g. velocity, we need to plot a graph in which the direction is fully specified by the angle of the vector relative to the co-ordinates. In fact, a vector can also be regarded as the set of all directed line segments that are equivalent to a given directed line segment. There is no sense of semblance here between the 'directedness' of powers and the direction of vectors. But even as a metaphor, the idea that powers are directed towards their effects is misleading. While in the M&A model the power-vector is 'directed' towards an existing point in the quality space, powers in fact have only the tendency towards the change of the state that their manifestation would yield. There is no existing *point* towards which they are directed.

9.4 Conclusion

The idea behind representing powers as vectors is that vectors can capture the key features of dispositional causes: directedness and intensity. As we have shown, the power-vectors in the M&A model are *not* vectors. In fact, we have shown that powers *cannot* be vectors. The idea is physically wrong and metaphysically confusing. If metaphysical views are to be consonant with current physical theories, then powers—whatever their merits—cannot be modelled as vectors. But then we asked: could it be that there is some sense in which powers are *analogous* to vectors, even if strictly speaking they are not vectors? We have shown that the analogy is misconceived. The notions of magnitude and directedness of power-vectors are problematic and incompatible with the idea that vector-powers operate within a quality space via vector addition. Powers cannot even be loosely analogous to vectors by means of their alleged intensity and directedness. It is, then, safe to conclude that the vectorial representation of powers is misleading and cannot cast any light on dispositional causation. Powers are not, and cannot be represented as, vectors.

Insofar as the new modal category of dispositionality is essentially dependent on thinking of powers as vectors—insofar as, that is, dispositionality is captured by the directedness of power-vectors—its plausibility as an intermediate kind of modality between necessity and contingency can be doubted.

Acknowledgements Many thanks to Anne Sophie Meincke for many useful comments on a previous draft. I (E. P.) would like to thank the Rotman Institute of Philosophy at Western University, USA, for funding my research via a CRC postdoctoral research fellowship, in the period 2014–2016 during which this chapter was co-authored.

References

Anjum, Rani L., and Stephen Mumford. 2018. *What Tends to Be. The Philosophy of Dispositional Modality*. London: Routledge.

———. This volume. Powers, Probability and Statistics. In *Dispositionalism. Perspectives from Metaphysics and the Philosophy of Science (Synthese Library* 417*)*, ed. A. S. Meincke, 115–130. Cham: Springer.

Geach, Peter T. 1961. Aquinas. In *Three Philosophers*, ed. G. E. M. Anscombe and P. T. Geach, 56–125. Oxford: Basil Blackwell.

Hitchcock, Christopher. 2007. What's Wrong with Neuron Diagrams? In *Causation and Explanation*, ed. J. K. Campbell, M. O'Rourke, and H. Silverstein, 69–92. Cambridge, MA/London: MIT Press.

Ioannidis, Stavros, and Stathis Psillos. 2019. Review of Rani Lill Anjum and Stephen Mumford. *What Tends to Be: The Philosophy of Dispositional Modality*. Routledge, 2018. *Notre Dame Philosophical Reviews*. 2019.03.35. https://ndpr.nd.edu/news/what-tends-to-be-the-philosophy-of-dispositional-modality/.

Lombard, Lawrence B. 1986. *Events: A Metaphysical Study*. London: Routledge/Kegan Paul.

Manley, David, and Ryan Wasserman. 2007. A Gradable Approach to Dispositions. *The Philosophical Quarterly* 57 (226): 68–75.

Martin, Charles B. 2008. *The Mind in Nature*. New York: Oxford University Press.

Mill, John S. 1882. *A System of Logic*. London: Parker.

Mumford, Stephen, and Rani L. Anjum. 2011a. *Getting Causes from Powers*. Oxford: Oxford University Press.

———. 2011b. Spoils to the Vector: How to Model Causes If You Are a Realist About Powers. *The Monist* 94 (1): 54–80.

Whitehead, Alfred N. [1929] 1978. *Process and Reality: An Essay in Cosmology*, corrected ed., D. R. Griffin, and D.W. Sherburne. New York: Free Press.

Chapter 10
A Dispositional Account of Causation, with Some Remarks on the Ontology of Dispositions

Alexander Bird

Abstract I outline and consider the prospects for (and significance of) an account of causation in terms of dispositions. The basic idea is (roughly) that C causes E when C is the stimulus of some disposition of which E is the manifestation. Such an account suggests that causes are conditionally sufficient (rather than necessary) for their effects. I consider what semantics for subjunctive and counterfactual conditionals this requires, and examine the consequences for making a distinction between causes and mere conditions and between collective or complete causes and mere parts of causes. Finally, I review the significance of this account of causation for a metaphysics of powers (potencies, essentially dispositional properties). I argue that it would be a mistake to see a successful dispositional account of causation (whether this or another account) as vindicating an ontology of powers.

Keywords Dispositions · Causation · Ontology · Conditionals · Transitivity of causation · Powers

10.1 Introduction—Dispositions and Causes

What is the relationship between dispositions and causes? One approach to understanding dispositions analyses them in terms of causes. Lewis (1997) supplies the paradigm of this approach. The basic idea is:

(CD) *o* is disposed to manifest *m* in response to stimulus *s* iff *o* and *s* together cause *m*.

Lewis tweaks this to avoid problems arising from the possibility of finks. But the basic idea is clear and plausible enough. Some philosophers (e.g. Jacobs 2007; Bird 2010; Mumford and Anjum 2010, 2011; Anjum and Mumford, Chap. 8, this volume), however, have sought to reverse the order of explanation, taking dispositionality to be more basic. (Mumford and Anjum take this to provide

A. Bird (✉)
Department of Philosophy, King's College London, London, UK
e-mail: alexander.bird@kcl.ac.uk

A. S. Meincke (ed.), *Dispositionalism*, Synthese Library 417,
https://doi.org/10.1007/978-3-030-28722-1_10

significant ontological lessons. I disagree, and I shall return to that issue towards the end of this chapter.)

A flammable liquid is ignited and catches light. The igniting of the liquid is the cause, and it is tempting to think that the flammability of the liquid plays some role here: it is because the liquid is flammable, i.e., disposed to catch light when ignited, that the igniting of the liquid can cause it to catch light. This suggests a simple dispositional analysis of causation that in effect reverses (CD):

(SD) *c* causes *e* iff for some object *o* and some disposition D, *o* possesses D and *c* is its stimulus and *e* is the manifestation of this instance of D.

Note that on this view it is the stimulus of the disposition (the igniting of the liquid) that is the cause, not the disposition itself (the flammability of the liquid). This differs from the approach of Mumford and Anjum (2010, 2011) who focus on the disposition as the cause. I hold that the disposition is not a cause but is a causal condition of the effect—I defend this distinction between causes and conditions in Sect. 10.4 below.

The strategy I pursue for assessing (SD) is as follows.

- I shall first assume that there is a simple relationship between dispositions and conditionals, (CA). This will allow me to investigate a proposition, (SD–CA)′, that is a consequence of (SD) *plus* (CA). (SD–CA)′ identifies causation with a certain kind of conditional, which I call *subjunctive sufficiency*. I show that the subjunctive sufficiency view has significant merits.
- I then look at the weaknesses of the subjunctive sufficiency view, i.e. of (SD–CA)′. I argue that the weaknesses in (SD–CA)′ are not due to (SD) but to (CA). I conclude that (SD) is a plausible account of causation: it has the strengths of (SD–CA)′ but can avoid its weaknesses.

This strategy should be seen as analogous to the investigation of a scientific theory (such as the kinetic theory of heat) by making certain simplifying assumptions (e.g. that molecules are point particles with no attractive forces between them). We first see that the theory plus simplifying assumptions deliver fairly accurate results in important cases, delivering a good degree of confirmation. We see also that the theory delivers inaccurate results in other cases (e.g. when gases are under high pressure). We then argue that the imperfections are due to the strict falsity of the simplifying assumptions, and that the inaccuracies diminish when the simplifying assumptions are replaced by more realistic ones. This delivers further confirmation. In our case using the simplifying assumption (CA) has two benefits. First, 'disposition' is to some degree a term of art, and our reasoning and intuitions are better developed with conditionals than with dispositions. Secondly, by discussing the dispositionalist proposal in terms of conditionals, we will be able to see more clearly the contrast between this view and Lewis's counterfactual conditional account of causation.

10.2 (SD) Plus the Simplifying Assumption—The Conditional Analysis of Dispositions

Let us turn then to the link between dispositions and conditionals. We'll start with the simple conditional analysis of dispositions:

> (CA) *o* is disposed to yield manifestation *m* in response to stimulus *s* iff were *o* to receive stimulus *s* it would yield manifestation *m*.

(CA) is false. But it isn't far from the truth. The standard counterexamples – finks, masks/antidotes, and mimics – require fairly unusual setups (Johnston 1992; Martin 1994; Bird 1998). (CA) is for many purposes a good approximation to the truth. As I explain later the standard counterexamples mentioned do not bear on the questions at issue in this chapter. Nonetheless we will consider other cases where (CA) fails and will see what implications those failures have for the current project.

Putting (SD) and (CA) together we have the following account of causation:

> (SD–CA) *c* causes *e* iff for some *o*, were *o* to receive stimulus *c*, then *o* would yield *e*, *c* occurs, and *e* occurs.

For now we may treat the existential quantification 'for some *o*' as playing only a dummy role. If the counterfactual '*c* occurs $\Box\!\!\rightarrow$ *e* occurs' is true, then the world as a whole is something such that were it to receive stimulus *c* it would yield manifestation *e*. (This is something to which we will return.) Consequently, we can simplify (SD–CA) to:

> (SD–CA)′ *c* causes *e* iff *c* occurs $\Box\!\!\rightarrow$ *e* occurs $\wedge$ *c* occurs $\wedge$ *e* occurs.

(SD–CA)′ needs careful explanation and defence, which it will receive in the following sections. For present, '$\Box\!\!\rightarrow$' symbolises, as is usual, a certain species of conditional whose instances are often counterfactual conditionals. But not all are; counterfactual conditional statements in English imply, pragmatically at least, the falsity of their antecedents (as their name suggests). But we do not take '$p \Box\!\!\rightarrow q$' to imply the falsity of p. The conditionals in question should preferably be called *subjunctive* conditionals, since that is the grammatical mood typically employed in the counterfactual case ('had you gone out now, you would have got wet') and the non-counterfactual case ('were you to go out now, you would get wet'). In what follows the relation symbolised by '$\Box\!\!\rightarrow$' will often be such a non-counterfactual subjunctive conditional.

(SD–CA)′ makes causes conditionally sufficient for their effect. '$\Box\!\!\rightarrow$' involves an implicit reference to background conditions that are relevantly similar to actual conditions. '*c* occurs $\Box\!\!\rightarrow$ *e* occurs', says that given such conditions, if *c* occurs then *e* occurs; that is, in such conditions, the occurrence of *c* suffices for the occurrence of *e*. This conditional sufficiency I call *subjunctive sufficiency*. Subjunctive sufficiency contrasts, of course, with Lewis's claim that causation amounts to counterfactual necessity. According to Lewis, causation between *c* and *e* requires '*c* does not occur $\Box\!\!\rightarrow$ *e* does not occur' (or its ancestral), i.e. causes are conditionally necessary for their effects.

In the remainder of this chapter (after an initial clarification regarding the semantics of '$\Box\!\!\rightarrow$') I shall develop (SD–CA)′ to see how far it can be taken as an account of causation. I shall then consider objections to (SD–CA)′. We will see that to a large degree these objections can be explained away as arising from the counterexamples to (CA). To that extent (SD) remains intact. I shall conclude by considering the extent to which (SD) provides an illuminating account of causation.

10.3 The Denial of Centering

One immediate problem with (SD–CA)′ is that it seems to have the consequence that any actual event is the cause of any other actual event. According to Lewis's account of counterfactuals the following holds:

$$\text{(C) } \mathrm{A} \wedge \mathrm{B} \rightarrow \mathrm{A} \mathbin{\Box\!\!\rightarrow} \mathrm{B}.$$

(C) and (SD–CA)′ entail:

$$(c \text{ occurs} \wedge e \text{ occurs}) \rightarrow c \text{ causes } e.$$

And so any two actual event are causally related.

(C) is the (strong) centering condition. In terms of Lewis's possible worlds semantics for counterfactuals, (C) is the claim that no world is as similar to the actual world as the actual world itself. A defender of (SD–CA)′ must reject (C).[1]

There are indeed good reasons to reject centering. In the current context, one is the fact that (C) and (CA) together entail that every two actual events are dispositionally related. One might regard that as further proof that (CA) is defective. But I think that would be a mistake. Consider two scientists discussing whether some experimental sheet of glass is fragile or not. A says, "The glass is fragile; if I were to strike it, it would shatter". "No," says, B, "This is really good stuff, I promise you. It isn't fragile; even if you were to strike it fairly hard, it wouldn't shatter". "Well then," says A, "Let's see." And so saying A takes up a hammer and strikes the sheet of glass with moderate force. Now consider two scenarios. In scenario I the hammer strikes the glass and the glass shatters. In scenario II, just as A is striking the glass, an enormous explosion in a neighbouring laboratory violently rocks the building, causing the glass to shatter a fraction of a second after the hammer makes contact with the glass.

[1]Rejecting centering does not entail rejecting weak centering, which is the denial that there is a possible world *more similar* to the actual world than the actual world itself. That said, reasons for rejecting centering might well lead us to reject weak centering also. See Gundersen 2004, p. 12 f.

We would regard scenario I as vindicating A's claims that the glass was fragile, and that it would shatter when struck. But we would not regard scenario II as vindicating A. Nonetheless, A, even in scenario II, can claim that according to (C) he was correct at least in saying that the glass would shatter if struck. But the naturalness of their conversation shows that such a response is mistaken. The explosion ruined their experiment. Nothing said by either A or B was confirmed or refuted in scenario II. In which case (C) is erroneous.

The outcome in scenario I is not an irrefutable confirmation of A's claims. After all, there *may* have been some hidden process in scenario I that operated like the explosion in scenario II to cause the glass to shatter quite independently of the striking. Since the scientists did not notice anything of the sort, they can regard the experiment as pretty strong confirmation of A's claims. But in analogous experiments we may not be so confident. Replication may reduce the probability that the outcome was not the result of the intervention but was instead the result of some hidden independent process. A control, if available, will often serve that purpose better. Had the scientists possessed a second piece of glass that was not struck, the fact that it shattered in scenario II would have shown them that the shattering of the struck glass tells them nothing about the truth of the subjunctive 'were it struck, it would break'. Correspondingly, such a control in scenario I (which does not break) would raise their confidence that the cause is the striking, not some other, hidden cause.

Before continuing I should address the worry that it is not (C) that is at the root of our concern, but (CA), which we already know to be faulty—thanks to mimickers and other things of that sort. Such a worry would have us drive a wedge between the two parts of A's claim, 'The glass is fragile' and 'if I were to strike it, it would shatter.' According to that worry, in scenario II the subjunctive 'if I were to strike it, it would shatter' is shown to be true; but because (CA) is false, it does not follow that the glass is fragile. As mentioned, the naturalness of the conversation between A and B and the unnaturalness of taking scenario II to vindicate the subjunctive claim, suggest that (CA) has more going for it intuitively than (C). But at this point I want to address the suspicion that the very same reasons we glean from a more detailed consideration of (CA) and which lead to its rejection, are at play here.

Consider scenario III, in which scientist A has attached a small explosive device to the same sheet of glass. That device has a very sensitive detonator. A strikes the glass, the striking disturbs the detonator, and the resulting explosion shatters the glass. In this case B is forced to concede, once she realises what has happened, that A's subjunctive 'if I were to strike it, it would shatter' has been verified. But B may reasonably reject A's claim that the glass is fragile. Thus scenario III does show that (CA) is false,[2] and at the very least needs modification. But notice the different role of mimicking in the two cases. Consider:

---

[2]At least if we assume: x is fragile $\equiv x$ is disposed to shatter when struck.

(a) c occurs $\wedge$ e occurs;
(b) c occurs $\Box\!\!\rightarrow$ e occurs;
(c) there is a disposition to manifest e in response to c.

So:
(a) $\wedge$ (C) entails (b);
(b) $\wedge$ (CA) entails (c).

In scenario III we have an entirely unobjectionable subjunctive conditional, but that conditional mimics the *disposition*. Whereas in scenario II, the conjunction of the striking and the (independently) shattering glass mimics the *conditional*. In III we thus accept the conditional but deny the disposition: the correct description of scenario III denies the inference of (c) from (b), i.e., denies the truth of (CA). In describing scenario II, we accept the conjunction but deny the conditional, i.e., we deny the inference of (b) from (a), i.e., we deny (C). Thus the falsity of (CA) plays no part in our rejection of (C).

Other philosophers have pointed to the more general counterintuitive consequences of (C). Alan Hájek (2007, pp. 46–48), after remarking that most people would be puzzled by, rather than assent to, counterfactuals joining unrelated actual facts, such as 'If Canberra were the capital of Australia then the moon would have large craters', goes on to point out that matters are worse when the events are related but in such a way that the antecedent reduces the chances of the consequent. Furthermore, since '$A \wedge B$' is symmetrical for A and B, $A \wedge B$ entails not only $A \Box\!\!\rightarrow B$, but also $B \Box\!\!\rightarrow A$. So both of the following are true of the 2000 U.S. presidential election: 'if Gore had won the popular vote, then Bush would have won the election' and 'if Bush had won the election, then Gore would have won the popular vote'.[3]

The rejection of centering means that when A and B are actually the case, more worlds are relevant to the truth of $A \Box\!\!\rightarrow B$ than just the actual world. Which additional worlds are these? Nozick (1981), who needs a subjunctive conditional that denies (C), takes the relevant worlds to be the set of *nearby* possible worlds, that is the set of worlds closer (more similar) to the actual world than some threshold distance. Gundersen (2004) takes the worlds to be those that are *normal*. Taking the latter route means the rejection of weak centering—since abnormal things do occur, the actual world may not be among the normal worlds in certain respects. Hence the set of worlds relevant to the truth of $A \Box\!\!\rightarrow B$ would not only include worlds other than the actual world, but need not include the actual world at all. One might of course regard the relevant set as intersection of the sets of nearby and normal worlds

[3]Hájek also discusses cases that turn on the incompatibility of 'If x had occurred, y might not have happened' and 'If x had occurred, y would have happened'. E.g., I am about to toss a fair coin. I say truly, 'if I were to toss this coin, it might land tails'. I do toss the coin and it lands heads. Thus, according to centering, 'if I were to toss this coin, it would land heads' would have been a correct thing to say. But the truth of the latter is incompatible with what I did say. For an extended discussion of centering and an alternative semantics for counterfactuals, based around *normality* rather than *similarity* and which rejects centering and weak centering, see Gundersen 2004.

(or one's conception of normality might imply nearness, even if not implied by it). For current purposes I shall take the *nearby* worlds to be relevant to the truth of subjunctive conditionals, although I shall also consider the benefits of adding the restriction to normal worlds.

10.4 Causes and Conditions

Lewis, along with many other philosophers who take counterfactual necessity as the mark of causation, draws no metaphysical distinction between a cause of an effect and a condition for that effect's occurrence. Yet, this is a distinction that non-philosophers find natural and easy to make.[4] In an ordinary case of a struck match lighting, every non-philosopher takes *the* cause of the lighting to be the striking and will consign the presence of oxygen to a lower status. It was the flicking of the switch that caused the bulb to illuminate, not the continued functioning of the power station supplying the electricity. The distinction between a cause and a causally necessary background condition is even more obvious the further back we trace such conditions. A person's birth is a necessary condition of their death at aged 82 of a heart attack, but their attempting to run a marathon was its cause.[5] Those who think that causes are counterfactually necessary conditions must regard all of these events as causes.

An advantage of the subjective sufficiency account (SD–CA)′, and of the dispositional account (SD) it approximates, is that, unlike the counterfactual necessity view, the view respects and explains the natural cause-condition distinction.

Lewis (1986, p. 162) holds the cause-condition distinction to be merely pragmatic—all the events mentioned above are indeed causes of the events for which they are necessary conditions. But our interests may lead us to focus our attention on one of the many causes and to pick it out as *the* cause:

> We sometimes single out one among all the causes of some event and call it "the" cause. Or we speak of the decisive or real or principal cause [...] I have nothing to say about these principles of invidious discrimination.

Likewise Hall (2004, p. 228) sermonises (his term),

> Suppose that my favorite analysis counts the big bang as among the causes of today's snowfall [...] How easy it is to refute me, by observing that if asked what *caused* the

---

[4]The advantage of this account in making the cause–condition distinction was made clear to me by reading Broadbent (2007a, 2008). Broadbent himself proposes a 'reverse counterfactual' account of causation, where the key counterfactual is: $\neg e \; \Box\!\!\rightarrow \neg c$ (see also Broadbent 2007b). This is a sufficiency account of causation and so, like mine, generates the cause–condition distinction.

[5]Those of Humean inclinations may not like this example, since birth is a logically necessary condition of death and so this may be thought to introduce an illegitimate necessary connection between distinct events. In which case we may substitute for a person's birth, their being fed as a child or their being administered penicillin during an early illness.

> snowfall (better still: what was *the* cause of it), we would never cite the big bang! Of course, the right response to this "refutation" is obvious: It conflates the transitive, egalitarian sense of "cause" with a much more restrictive sense (no doubt greatly infected with pragmatics) that places heavy weight on salience.

Hall's response is disingenuous. What reason do we have for supposing that there is an egalitarian sense of cause that encompasses the Big Bang as a cause of every particular event, and my grandmother's birth as a cause of all my actions, and so forth—except for the fact that counterfactual accounts, such as those of Lewis and Hall, deliver that result? Elsewhere our intuitive reactions to certain cases are regarded as data that, if possible, ought to be accommodated by a satisfactory theory. But here those reactions are dismissed as infected by pragmatics.

The resort to pragmatics is an insufficient response. For a start, as Broadbent (2008) rightly complains, those who make it never offer a substantive account of the pragmatic principles at work. And when we turn to our best general account of the pragmatics of discourse, Grice's account of conversational implicature, we find that Grice's principles do not deliver the result that is required, as Menzies (2004, p. 147) points out.[6] Moreover, salience and pragmatic concern may well focus one's interest on a necessary condition without thereby elevating it to the status of cause. A speeding motorist causes an accident. The cause of the accident is clearly their excessive speed. Even so, one might take an interest in other factors. The town council may conclude that any of a variety of speed reduction devices (speed humps, road narrowing, speed cameras, etc.) would have prevented the accident. But that does not mean that the council concedes that the lack of humps, the width of the road, or the absence of speed cameras are each causes of the accident. In which case the application of the honorific 'cause' is not correlated with the focus of our interests.

For a contrasting account of the cause–condition distinction, let us now turn to (SD–CA)′. Given our dropping of centering, this requires that the material conditional, $c \rightarrow e$, holds in nearby worlds as well as in the actual world. Consider $c \equiv$ the match is struck, and $e \equiv$ the match lights. In some nearby worlds the match may not be struck, in which case $c \rightarrow e$ is true. In others it is struck, but since in nearby worlds oxygen is present the match lights, so again $c \rightarrow e$ is true. The striking of the match causes its lighting. But now consider $c' \equiv$ oxygen is present. In some nearby worlds the match is not struck, but oxygen is still present. Hence in those worlds $c' \rightarrow e$ is false, and so $c' \,\Box\!\!\rightarrow e$ is also false. The presence of oxygen is not a cause of the lighting. Likewise, *mutatis mutandis*, for the other cases.

(SD–CA)′ is being employed as an approximation to the dispositional account (SD). We can also see directly that the latter respects the cause-condition distinction. Consider the lighting match. Clearly the unstruck match has a disposition to light in response to the stimulus of being struck. But it does not have the disposition to light in response to the stimulus of being in the presence of oxygen—not under these circumstances. The elderly gentleman was disposed to die of a heart

[6]Menzies uses this to motivate a contextual account that retains difference-making.

attack in response to excessive exertion, but not disposed to die of a heart attack in response to being born (or being fed as a child etc.).

Lewis in effect has an error theory of our normal causal talk, since '*the* cause' implies only one cause, whereas according to Lewis there is never only one cause. While such an error theory could be right, without a satisfactory account of the pragmatics at work, it lacks an explanation of why we make that error. The dispositional view of causation, however, because it respects our intuitive distinction between cause and condition, has a significant advantage.

10.5 Collective Causes

The dispositional account has the following feature. If two uncommon events coincide to cause an effect, so that both are required—neither is sufficient in the absence of the other—then neither counts as a cause of the effect. For example, a fire occurs because a fuel line leaked and a build-up of static electricity (perhaps due to unusual weather conditions) caused a spark. The setup was not disposed to bring about a fire either in response to the fuel leak or in response to the spark, but only in response to both together. So neither individually causes the fire. In counterfactual terms, [there is a leak $\Box\!\!\rightarrow$ there is a fire] is not true, since in a nearby world there is no spark and so no fire, and likewise [there is a spark $\Box\!\!\rightarrow$ there is a fire] is not true since in a nearby world there is no fuel leak and so, again, no fire.

This might be thought to be a disadvantage of the dispositional account, but I shall argue the opposite—our pre-theoretic judgments favour what I call *collective* causes: causes whose parts are not individually causes. I suggest that philosophers may be inclined to think that the fuel leak and the spark are each causes because of so lengthy an exposure to the Hume–Lewis negative counterfactual view which make any necessary condition a cause. But as we have already seen, everyday causal talk is not so profligate with ascriptions of cause, making a distinction, as it does, between causes and mere conditions. In this case too, everyday causal talk is disposed *not* to take the individual events as causes but rather takes them together as a collective cause. In answer to the question, 'what caused the fire?' it would generally be regarded as incorrect, rather than at worst misleadingly incomplete, to say just 'the fuel leak'. No, what caused the fire was the *co-occurrence* of the fuel leak and the spark. This answer is most clearly correct when the two (or more) components of the cause are identical. What caused Oedipus to be blind? Not that he removed his left eye, nor that he removed his right eye, but that he removed *both* eyes. A twin prop aeroplane crashes as a result of the failure of both of its two engines: the cause of the crash is just that, the failure of both engines, not the failure of either one of them. I fill my tank with 50 l. of fuel so as not to have to stop on my car journey. The first 25 l. that I put into the tank does not enable ('enable' = 'causes to be possible') me to get to my destination, nor does the second 25 l., only that I filled with 50 l. Fraser MacBride (2005), in a rather different context, also appeals to the idea of collective causation, in his example of an unfortunate man stung to death by a swarm of bees:

no individual bee sting is the cause of his death; rather the many stings are collectively the cause of death.[7]

Again, the fact that the dispositional account, both as (SD–CA)′ and as (SD), explains the pre-theoretic verdicts, but the counterfactual necessity view does not, favours the former. (SD–CA)′ tells us:

> [there is a fuel leak □→ there is a fire] is false;
> [there is a spark □→ there is a fire] is false; but
> [there is a fuel leak and there is a spark □→ there is a fire] is true.

And likewise,

> [the left engine fails □→ the aeroplane crashes] is false;
> [the right engine fails □→ the aeroplane crashes] is false; but
> [the left engine fails and the right engine fails □→ the aeroplane crashes] is true, etc.

Similarly, the first 25 l. and the second 25 l. may both be individually necessary for my travelling without needing to stop, but only the full 50 l. is sufficient. Thinking about the fire directly in dispositional terms, the state of affairs before the fire was disposed to bring about a fire as a result of the co-occurrence of a fuel leak and spark, but was not disposed to bring about a fire in response to either of those events individually. The plane was disposed to crash if both its engines failed, but not disposed to crash if just one failed.

10.6 Counterfactual Dependence?

Ned Hall (2004) claims that there are two concepts of causation, which he calls *dependence* and *production*. Dependence is negative counterfactual dependence—counterfactual necessity. Hall offers a tentative analysis of production, a concept that is closer to some kind of sufficiency. In Hall's view, to account for all our causal ascriptions we need both of these distinct concepts. I suggest that we can do without dependence and that production should be understood in terms of counterfactual sufficiency.

Hall (2004, p. 253) considers someone who might react to his discussion by denying that counterfactual dependence is causation. The defender of counterfactual dependence says,

> Nonsense; counterfactual dependence *is too* causation. Here we have two wholly distinct events; moreover, if the first had not happened, then the second would not have happened. So we can say—notice how smoothly the words glide off the tongue—that it is in part *because* the first happened that the second happened, that the first event is partly *responsible for* the second event, that the occurrence of the first event helps to *explain why* the second event happened, and so on. (Emphases in original).

Note that Hall, says 'in part' and 'partly' and 'helps'. So it seems that even Hall is uncomfortable with saying *a* causes *b* when *a* is just one among several necessary

[7]MacBride is concerned to show that causation is a multigrade relation.

conditions (and hence not sufficient). Indeed, everything Hall says above, except the first sentence, is consistent with the denial that the first event causes the second event. The best explanation of the natural use of the qualifiers 'partly' and the like is that the events in question are *parts* of causes—how else should we interpret '*b* partly because *a*' except as 'for some *c*, *b* because *c* and *a* is part of *c*'?

Since I reject dependence as a component of an analysis of causation, I am under some obligation to offer an explanation of why it should seem to many that it is a (or *the*) central feature of causation. The fact that dependence will pick out parts of causes is itself one reason why we might mistakenly fix on dependence as an account of causation. If dependence or an account based on it (such as Lewis's) were, by the lights of its supporters, to be a satisfactory account, then it would differ in is evaluations from the dispositional view only insofar as it also counts as causes (a) parts of causes and (b) conditions. We have discussed conditions above. Why theorists have been willing to swallow this counterintuitive aspect of the dependence view is in part testament to the power of theory to inform our or override our intuitive judgments. This may be helped by misleading examples that seem to suggest that what we regard as a cause and what we regard as a condition is interest relative. So while we may think that the presence of oxygen is a mere condition for the fire, not a cause, we may be persuaded to think otherwise by relating a story in which a seemingly similar set of events occurs in which we undoubtedly do identify the oxygen as a cause. If oxygen were normally absent from some process, but was accidentally introduced and as a result a fire occurred, then the oxygen would be a cause. But that does not show that it was a cause in the case where it *is* normally present. The two cases differ with respect to their dispositions and to the truth of the positive subjective conditionals, and so the difference in our intuitive judgment is not explained by a difference in our interests.

Another reason why dependence seems important is that it is related to responsibility. Human nature being what it is, we are very interested in who can be responsible for those occurrences that we would rather had not happened—we are more interested in that than in finding out who brought about events that we are happy with. As such responsibility is a matter of necessary conditions—one is responsible for what one could have prevented but failed to. And it may be natural—although strictly false—to think that what we are doing by identifying those persons and actions that are responsible for a certain outcome is identifying the cause of the unfortunate events in question. In which case we will equally naturally associate causation and necessary conditions, and given that association we can generalise beyond cases of human action and responsibility to relations between events in general.

Persons and actions that are responsible for certain outcomes will often be causes of them. This is because intentional behaviour is typically both a necessary *and* a sufficient condition of the intended outcome. Mary wants her garden to be watered. On a summer's day she will want to bring about that outcome, i.e. cause it, and so engage in an action that is subjunctively sufficient for the garden to be watered. On the other hand, if the desired outcome is going to happen anyway, and we know that, then we have no reason to act, and human nature being what it is we typically desist from acting. If Mary can see that it is about to rain, she won't go to the effort of getting the hose out. So our actions are typically not only sufficient conditions of their outcomes but also necessary conditions—we often do not act unless it is necessary to do so. This means that actions that are causes of their effects are not only sufficient but are often also necessary conditions of those effects and since they correlate it is not surprising that responsibility, causation, and necessary conditions are held to be intimately related. But causation and responsibility can come apart, as cases of negligence show. A landlord's negligently failing to check the safety of his building may be a necessary condition, but not a sufficient one, of the fire that broke out. That fact may entitle us to hold him responsible, but we do not have to regard his negligence as a cause of the fire.

10.7 Problems and Solutions

Having given a sketch of a dispositional account of causation above, I shall now turn to potential objections. The discussion so far has concentrated on subjunctive sufficiency, which—given (CA)—is a consequence of (SD). But, as remarked, (CA) is false. It is approximately true, but in some cases dispositions and subjunctive conditionals come apart. Given that they do come apart, we may ask, which of (SD) and (SD–CA)′ is true. As we shall see, certain objections to (SD–CA)′ disappear once we focus on (SD). The account being proposed here is thus a dispositional rather than subjunctive account.

(a) If **a** were crimson, **a** would be red, and **a** is in fact crimson. But we do not regard **a**'s being crimson as a cause of its being red. Counterfactual dependency views of causation have a similar problem: if **a** were not red, it would not be crimson, which, according to such views, suggests that being red is a cause of being crimson (which is even less plausible than being crimson causing being red). However, the dispositional view can avoid this problem by remarking on the difference between dispositions and counterfactuals. We do not think that everything is such that it has the *disposition* to be red if it is crimson. So such a case already constitutes a counterexample to the counterfactual analysis of dispositions, (CA). Rejecting (CA) and so (SD–CA)′ but retaining (SD) allows one to say that although it is true that if **a** were crimson, **a** would be red and that **a** is in fact crimson and red, there is no causation here because there is no disposition—this is one of the cases where dispositions and subjunctive conditionals come apart.

(b) Let it be the case that in the circumstances the only event that could cause *e* is *c* and that *c* did cause *e*. One might be inclined to say that it is true to say that were it that case that *e* occurs, then it is also the case that *c* occurs. Thus our subjunctive conditional account has the consequence that not only does *c* cause *e* but also that *e* causes *c*. Note that the counterfactual dependence view also suffers from this problem. Lewis's answer is to outlaw backtracking counterfactuals. If that answer is a good one, with independent motivation, then the dispositional view can use it also. However, there are doubts as to whether the rejection of backtracking counterfactuals is legitimate (especially as Lewis's own small miracles employ them). The dispositional account has an independent answer that does not appeal to the denial of backtracking,. In normal cases there will be a disposition to yield *e* in response to *c* without there being a disposition to yield *c* in response to *e*. (I am not saying that we must always rule out backwards directed dispositions, just that generally they do not occur, which suffices to make the distinction in the direction of causation that such cases require.)

(c) If *c* is subjunctively sufficient for *e* then $c + d$ is subjunctively sufficient for *e*, where *d* is causally independent of *e*. So spurious events my seem to be parts of causes where they are not. This is the rough analogue of the problem for Hempel that any condition can be added to laws and conditions that entail the explanandum without removing that entailment. So we get spurious explanatory and spurious causal factors. Again the appeal to dispositions removes the problem, since a fragile vase has the disposition to break in response to being stressed, but it does not have the disposition to break in response to the complex stimulus [being stressed and Canberra is the capital of Australia].

Objections (a)–(c) reinforce (SD) by showing that objections against (SD–CA)′ occur precisely where (CA) is false and so where (SD) and (SD–CA)′ come apart. These are exceptions that prove the rule, and confirm (SD) in the way that observations in science support an underlying theory when we find that although they are inconsistent with a simple model built on that theory they are consistent with more sophisticated models constructed on the same basic theory (for example, in Newtonian mechanics, observations of the motion of the moon refuted his own model of the moon's motion, but were found eventually to be in conformity with a mathematically more sophisticated model. Those observations thus confirmed the underlying Newtonian gravitational theory.)

It may be noted however that none of the objections considered so far relates to the standard counterexamples to (CA), viz. finks, antidotes/masks, and mimics. Whereas Mumford and Anjum (2011) motivate their dispositional account through the possibility of interference.

We can see why the standard counterexamples are not relevant. Fink and antidote cases occur when there is a disposition that receives its characteristic stimulus but because the disposition is removed or interfered with it does not produce its manifestation. So those are cases where an event (the non-actual manifestation) does not occur. Whereas, what we need in order to test an account of causation are cases of two *actual* events—the question is whether the account correctly classifies them as causally related or not. More relevant, then, are cases of mimics and the finkish production of a disposition. In the former case a cause brings about an effect in a way that mimics a disposition that is not present. For example, a robust iron pot is attached to a powerful bomb and a sensitive detonator. Strike the pot, the bomb explodes, and the pot shatters—mimicking fragility. Such a case is not any problem for (SD–CA)′ or (SD) in particular. In another kind of case one event both brings a disposition into existence (which was not present when the event occurred) and also is the stimulus of the disposition, so bringing about its manifestation. (CA) wrongly suggests that the disposition in question was present when the initiating event occurred. In Martin's (1994) example, an electro-fink device is attached to a dead wire, but makes it live if a conductor touches the wire—the conductor thus experiences a current. So the wire behaves as if it were live at the moment of touching even though it is not. Although the wire is not live, we can say that the *overall* setup (wire plus electro-fink) is disposed to conduct a current to a conductor that touches it. And so we are able to get the result that (SD) predicts.

I noted that Mumford and Anjum (2011) motivate their version of a dispositional account of causation in part through the possibilities of interference: causes can always be interfered with and so cannot necessitate their effects. So the relevant property of the cause (which for Mumford and Anjum is the disposition, not its stimulus) must be something weaker than necessitation, something they call the dispositional modality (which they argue is the most basic form of modality). In contrast, I have just argued that the possibility of interference is not itself a reason to adopt a dispositional account of causation in preference to the subjunctive sufficiency view (though I have given other reasons for that preference).

Whether causes cannot necessitate their effects for this reason depends on what one takes 'necessitation' to be. Certainly, causes do not metaphysically necessitate their effects (there are metaphysically possible worlds with the cause but without the effect); nor do they nomically necessitate their effects (there are nomically possible worlds with the cause but without the effect). But this does not show that there is no plausible conception of necessitation such that causes necessitate their effects. Subjunctive sufficiency is a kind of necessitation, but much weaker and so does not require that causation rule out the metaphysical or nomic possibility of interference.[8] Because of the denial of centering, subjunctive sufficiency does rule out the possibility of interfering in the nearest possible worlds. And that is what we want from causation. If igniting the liquid caused it to catch fire, then it will be the case that were we to ignite the liquid in circumstances that are identical in relevant respects then it catches fire in those circumstances too.

[8]Mumford and Anjum (2011, p. 57) might decline to regard subjunctive sufficiency as a kind of necessitation, since it does not pass their antecedent strengthening test: if A necessitates B, then A + φ necessitates B (for any φ). But as Lowe (2012) argues, this is too strong a condition on necessitation.

10.8 Ontological Considerations (I)—Complex Dispositions and the Non-transitivity of Causation

Dispositions are often made of combinations of simpler dispositions. My alarm clock is disposed to ring when the hour hand reaches a certain point. Clearly the latter event is the stimulus to the first of a series of dispositions, the last of which manifests itself in the ringing of the clock. Do we always have a disposition when we can find a sequence of dispositions like this? It is not clear that we do. A strong gust of wind causes snow to dislodge from a precipice. That initiates an avalanche, the avalanche flattens a tree in its path. In each case we may identify a disposition: the precariously balanced snow drift was disposed to fall in response to a gust of wind; the accumulated snow in the valley was disposed to become an avalanche in response to a disturbance, such as the falling snow drift. The avalanche was disposed to flatten trees in its path. Should we say that something (the valley?) was disposed to flatten the tree in response to the gust of wind? That does seem to stretch things, to say the least. Hitherto I have said nothing on the topic, except to accept that we can be entirely liberal in ascribing dispositions. But that assumption has not played any role in the arguments above. Here I want to consider the advantages of being somewhat more restrictive in our ascriptions of dispositions, restrictive in a way that is in tune with our disinclination to ascribe an encompassing complex disposition in the valley case.

Note first that our disinclination in that case mirrors our disinclination to ascribe a causal relation. We are *not* inclined to say that the gust of wind caused the tree to be flattened. Now, once again, that could be ascribed to pragmatics. But the explanation I suggest here is the relationship between causation and dispositions given by (SD). It does not matter that we do not absolutely reject the ascription of causation or of a disposition—both concepts are vague. What does matter is that the degree of disinclination is correlated, which confirms (SD).

The fact that we do not endorse immediately the causal connection suggests that our concept of causation is not such that we take it to be obviously transitive. Lewis, in defining causation as the ancestral of the counterfactual dependency relation, makes causation essentially transitive. That is despite the existence of counterexamples to the transitivity of causation. The dispositional account, if it is not entirely liberal with its ascription of dispositions, will also reject the transitivity of causation. Here is a characteristic example: Kvart suffers an industrial accident in which his finger is severed. He rushes to the surgeon who skilfully reattaches it so that a year later the finger is as healthy as it ever was. The accident causes Kvart to have surgery, the surgery causes the finger to be healthy at the later date, but the accident does not cause the finger to be healthy at that date.[9]

[9]This example is attributed by Hall (2000) to Kvart in connection with Kvart (1991). Some may find that such examples seem not to be examples of transitivity since the effects show no difference from the original state. But I think we get the same result even if we make changes to the final state. Imagine that the surgeon finds a small cyst and removes that while reattaching Kvart's finger. So the finger ends up in a healthier state than it was originally. Still, it is not the case that the accident caused the finger to be healthier than before.

We can see that such examples substantiate the explanation of the failure of transitivity that I propose: the fact that dispositions do not always concatenate into more complex dispositions, although they sometimes do. So although Kvart is disposed to get surgery for a severed finger and the surgeon is disposed to repair a severed finger with skill such that it is healthy a year later, we do not think that Kvart has the disposition to have a healthy finger in response to that finger being severed. Nor would we attribute any such disposition to the sum of Kvart and the surgeon nor any other entity.

It is natural to ask then: when do dispositions concatenate to produce a complex disposition and when do they not? It is not possible to give a simple answer here. But here are some pointers:

- This is an analogue of the special composition question for objects. Not any set of objects compose to form a further object, but sometimes they do. But finding the correct principle or principles of composition is highly non-trivial. Note that the vagueness surrounding composition of dispositions is not a strong objection to the idea that they do sometimes concatenate but not always: the same issue arises for the composition of objects, but should not drive us to either unrestricted composition (every set of objects forms a composed object) nor to nihilism (no non-singleton non-empty set of objects ever forms a composed object).
- Easy cases of combination are where we can identify a whole with a function which is realised by the component dispositions. The functions may be natural or artificial. For example, the cardio-vascular system is disposed to transfer oxygen from the air we breathe to the muscles and other organs. That disposition is made up of the disposition of the lungs to dissolve oxygen from the air in the blood stream and the disposition of the heart to pump the blood to the muscles. On the other hand, some combinations of dispositions are clearly too accidental to form a further disposition, as in the cases discussed.
- One area to look to for further insights is the literature on mechanisms. Similar issues arise for the ontology of complex mechanisms as for complex objects and complex dispositions. A conjecture worth considering is the following: the complex dispositions are those combinations of dispositions that can be said to form mechanisms.

10.9 Ontological Considerations (II)—Powers

I have suggested that the ontology of dispositions is not entirely trivial. It is not as simple as saying that there is a disposition for every possible pair of stimulus and manifestation conditions. On the other hand, that fact tells us nothing about what dispositions are. For all that has been said in the preceding section, dispositional properties may all or mostly be abundant properties—not genuine, ontic properties at all. Or they might be genuine properties, but not dispositional in nature or essence. Can we draw any conclusions from a dispositional account of causation regarding the nature of dispositions themselves?

Mumford and Anjum (2010, 2011) also offer a dispositional account of causation. They have a particular aim in their account, which is to vindicate an ontology of powers. The latter claims that properties (some at least) are dispositional in nature or essence. This contrasts with the view that no properties are by nature dispositional; rather a property might contingently have a dispositional character, but that character is conferred upon it by the contingent laws of nature—under different laws the same property might have had different effect and so a different dispositional character. Mumford and Anjum take the success of a dispositional account of causation to provide evidence in favour of the ontology of powers. So does the success of the current account (or indeed theirs) provide any such reason?

I do not think so. There is no clear inference from:

(A) All causation is the manifestation of some disposition.

to:

(B) Properties have a dispositional nature/essence.

Proposition (A) is consistent with the claim that the disposition that is associated with a given property depends on what laws of nature there are. The correctness (or otherwise) of this dispositional account of causation, (A), turns on facts about causal and dispositional relations in the actual world or worlds very much like it. Whether (B) is correct turns on the possibility (or not) of remote possible worlds with different laws and different dispositions associated with the same properties. So it is difficult to see how (A) could have any bearing on (B). The existence of powers is one way the world could come to have dispositionality and causation. Another way for there to be dispositionality and causation is for there to be contingent laws relating properties that are not powers (essentially categorical properties). The success of a dispositional account of causation (whether this one or that of Mumford and Anjum) is no better evidence for the former than for the latter. I suspect that any thought that such success supports the ontology of powers arises from a conflation of the concepts of 'power' and of 'disposition'. Indeed Mumford and Anjum (2011, p. 4) explicitly say that they use these terms interchangeably. At best that is misleading and at worst begs the question. For on that equation a dispositional account of causation is by definition a powers account of causation. But that rules out by fiat the possibility that a non-powers ontology can equally well account for both causation and dispositionality. (I expand at length on these ontological issues in (Bird 2016)).

10.10 Conclusion—Causation and Explanation

The following three facts have wide currency among philosophers, in the sense that each individually is held to plausible by many philosophers:

(i) Explanations are supposed, optimally at least, to provide sufficient conditions for their explananda. An explanation is often supposed to show why we could (armed with the explanation) expect the explanandum. Which it could hardly do if it failed to provide sufficient conditions. Hempel's DN model requires a non-elliptical explanation to be such that the explanandum is deducible from the explanans—hence the explanans is sufficient for the explanandum. An explanation may provide conditionally sufficient conditions for its explanandum, i.e., conditions that are sufficient given certain background conditions. An elliptical DN explanation is of this sort.

(ii) Causes are counterfactually necessary conditions of their effects; they are typically not sufficient conditions. This is the thrust of the Hume–Lewis counterfactual account of causation, according to which causation is understood in terms of counterfactual dependence: had the cause not occurred, the effect would not have occurred. Something can fulfil this condition and so be a cause without being such that it is sufficient for the relevant effect.

(iii) To identify a cause of an event is to provide an explanation of it. Even if causal explanations are not the only explanations there are, they do form one prominent species of explanation.

There is a tension among these claims. If causal explanations cite causes as explanations of effects, as (iii) tells us, then, according to (i), causes should then be at least conditionally sufficient for their effects. But according to (ii) causes are conditionally necessary, not conditionally sufficient for their effects.

One might suspect (i), holding that the theory of explanation is not in very good shape. After all, has not Hempel's account been roundly refuted? Yes, but note that the principal and most powerful objections to Hempel's model are directed against one direction of the equivalence of explanation and sufficient conditions: the claim that sufficient conditions for *e* always provide an explanation of *e*. Achinstein's (1983) poisoning case, for example, describes conditions that suffice for Jones's death (his ingesting a pound of arsenic) but which do not explain his death (because he was in fact killed by a bus). We are obliged by such examples to conclude that sufficient conditions do not always provide explanations. That leaves intact the view that to provide an explanation is to provide sufficient conditions in some form or other; so (i) survives.

In this chapter I have suggested that it is (ii) that is at fault. Causes are in a certain sense sufficient for their effects. For much of the discussion I supposed that a kind of conditional sufficiency, subjunctive sufficiency, provided the answer.

As a consequence of the focus on subjunctive sufficiency, we have learned that the entirely natural distinction between cause and condition is an entirely legitimate one, metaphysically speaking. That in turn leads me to suggest that in some cases several events may be part of one collective cause (each part is not itself the cause but is part of the cause).

As we have seen, subjunctive sufficiency is in fact, for several reasons, only a proxy for the relation of a stimulus of a disposition to its manifestation. Focussing on cases where dispositions and subjunctive sufficiency come apart allow us to avoid objections to the subjunctive sufficiency view of causation. Furthermore, thinking about the ontology of dispositions shows us why counterexamples to the transitivity of causation occur. But they should not be expected to reveal anything about other fundamental questions of ontology, viz. whether properties are powers.

Acknowledgements Research for this chapter was funded by the Arts and Humanities Research Council (Research Project AH/D503833/1). I am grateful also to Anne Sophie Meincke and to audiences in Oxford, Birmingham, and Geneva for comments on earlier versions.

References

Achinstein, Peter. 1983. *The Nature of Explanation*. Oxford: Oxford University Press.

Anjum, Rani L., and Stephen Mumford. This volume. Powers, Probability and Statistics. In *Dispositionalism. Perspectives from Metaphysics and the Philosophy of Science (Synthese Library* 417*)*, ed. A. S. Meincke, 115–130. Cham: Springer.

Bird, Alexander. 1998. Dispositions and Antidotes. *The Philosophical Quarterly* 48: 227–234.

———. 2010. Causation and the Manifestation of Powers. In *The Metaphysics of Powers. Their Grounding and Their Manifestations*, ed. A. Marmodoro. London: Routledge.

———. 2016. Overpowering: How the Powers Ontology has Over-Reached Itself. *Mind* 125: 341–383.

Broadbent, Alex. 2007a. *A Reverse Counterfactual Analysis of Causation*. Ph.D. thesis, University of Cambridge.

———. 2007b. Reversing the Counterfactual Analysis of Causation. *International Journal of Philosophical Studies* 15: 169–189.

———. 2008. The Difference Between Cause and Condition. *Proceedings of the Aristotelian Society* 108: 355–364.

Gundersen, Lars B. 2004. Outline of a New Semantics for Counterfactuals. *Pacific Philosophical Quarterly* 85: 1–20.

Hájek, Alan. 2007. *Most Counterfactuals Are False*. Unpublished article available at https://docplayer.net/32738-Most-counterfactuals-are-false-alan-hajek.html. Last accessed on 24 Oct 2018.

Hall, Ned. 2000. Causation and the Price of Transitivity. *Journal of Philosophy* 97: 198–222.

———. 2004. Two Concepts of Causation. In *Causation and Counterfactuals*, ed. J. Collins, N. Hall, and L.A. Paul, 181–276. Cambridge, MA: MIT Press.

Jacobs, Jonathan D. 2007. *Causal Powers: A Neo-Aristotelian Metaphysic*. Ph.D. thesis, Indiana University.

Johnston, Mark. 1992. How to Speak of the Colors. *Philosophical Studies* 68: 221–263.

Kvart, Igal. 1991. Transitivity and the Preemption of Causal Relevance. *Philosophical Studies* 64: 125–160.

Lewis, David K. 1986. *Philosophical Papers, Vol. II*. New York: Oxford University Press.

———. 1997. Finkish Dispositions. *The Philosophical Quarterly* 47: 143–158.

Lowe, E. Jonathan. 2012. Mumford and Anjum on Causal Necessitarianism and Antecedent Strengthening. *Analysis* 72: 731–735.

MacBride, Fraser. 2005. The Particular–Universal Distinction: A Dogma of Metaphysics. *Mind* 114: 565–624.

Martin, Charles B. 1994. Dispositions and Conditionals. *The Philosophical Quarterly* 44: 1–8.

Menzies, Peter. 2004. Difference-Making in Context. In *Causation and Counterfactuals*, ed. J. Collins, N. Hall, and L. Paul. Cambridge, MA: MIT Press.

Mumford, Stephen, and Rani L. Anjum. 2010. A Powerful Theory of Causation. In *The Metaphysics of Powers: Their Groundings and their Manifestations*, ed. A. Marmodoro. London: Routledge.

———. 2011. *Getting Causes from Powers*. Oxford: Oxford University Press.

Nozick, Robert. 1981. *Philosophical Explanations*. Cambridge, MA: Harvard University Press.

Chapter 11
Powers, Dispositions and Laws of Nature

Max Kistler

Abstract Metaphysics should follow science in postulating laws alongside properties. I defend this claim against the claim that natural properties conceived as powers make laws of nature redundant. Natural properties can be construed in a "thin" or a "thick" way. If one attributes a property in the *thin* sense to an object, this attribution does not conceptually determine which other properties the object possesses. The thin construal is underlying the scientific strategy for understanding nature piecemeal. Science explains phenomena by cutting reality conceptually in properties attributed to space-time points, where these properties are conceived of independently of each other, to explore then, in a separate step, how the properties are related to each other; those determination relations between properties are laws. This is compatible with the thesis that laws are metaphysically necessary. According to the *thick* conception, a property *contains* all its dependency relations to other properties. The dependency relationships between properties (which appear as laws in the thin conception) are *parts* of the properties they relate. There are several reasons to resist the thick conception of properties. It makes simple properties "holistic", in the sense that each property contains many other properties as parts. It cannot account for the fact that properties constrain each other's identity; it can neither explain why natural properties are linked to a unique set of dispositions, nor why and how this set is structured nor why the truth-maker of many disposition attributions is relational although the disposition is grounded on a monadic property.

Keywords Law of nature · Property · Natural property · Power · Disposition · Metaphysics · Necessity · Natural necessity · Determination

---

M. Kistler (✉)
Institut d'Histoire et de Philosophie des Sciences et des Techniques, Université Paris 1 Panthéon-Sorbonne, CNRS, Paris, France
e-mail: mkistler@univ-paris1.fr

A. S. Meincke (ed.), *Dispositionalism*, Synthese Library 417,
https://doi.org/10.1007/978-3-030-28722-1_11

11.1 Introduction

Many sciences seem to discover laws. Physics is full of examples: Kepler's laws, Newton's laws, Ohm's law, Maxwell's laws. Of course, the fact that certain generalisations or mathematical equations are called "laws" does not establish that they really express laws. Some generalisations, such as the Titius-Bode law, are called "laws" although everyone agrees that they do not really express a law, but only an accidental regularity. According to a widely shared conviction, statements such as Ohm's law express laws because they satisfy a certain number of conditions that set them apart from accidental universal statements. Let us accept as a working hypothesis that laws are expressed by equations or universal generalisations that can be used to explain and predict phenomena (or other laws that can themselves be thus used) and to justify counterfactual conditionals.[1]

Until fairly recently, the main philosophical debate on laws of nature opposed empiricists and nomological realists. For empiricists, laws are a special sort of regularities concerning events or facts, or supervene on such regularities. According to Lewis's version of the so-called Mill-Ramsey-Lewis account of laws of nature, "a contingent generalisation is a *law of nature* if and only if it appears as a theorem (or axiom) in each of the true deductive systems that achieves a best combination of simplicity and strength" (Lewis 1973, p. 73; italics Lewis's).[2] According to the main opposite view, nomological realism, laws are objective features of reality. Their existence depends neither on the existence of theories nor on the existence of scientists and subjects asking for explanations and predictions. According to the most influential version of nomological realism, developed by Dretske (1977), Tooley (1977) and Armstrong (1983), laws are necessitation relations between universals. On both traditional accounts, laws are contingent. This fits well with the intuition that the laws could have been different from what they are. However, such intuitions are not always trustworthy. It might well be that the contingency of laws seems intuitively convincing only to someone who confuses the fact that laws are discovered a posteriori (or are "epistemically contingent") with the fact that they are metaphysically contingent. Shoemaker (1984, 1998) and others have made a case for the opposite thesis, according to which laws are metaphysically necessary.[3] As I will show later, Shoemaker's reasoning leads to the result that laws are *conditionally* necessary, in the sense that all laws featuring property P exist in every world in which P exists. Against this, Mumford (2004) has argued that the premises of the

[1]It is notoriously difficult to provide a non-circular analysis of the difference between laws and accidental generalisations. See Goodman (1955). There are many proposals for adding more necessary conditions to those mentioned in the text, but they are all controversial. Bird (2006, p. 451) suggests, e.g., that laws must have some degree of fundamentality and must correspond to a new discovery, i.e. not be deducible from already known laws. This is not the place to enter the debate about what exactly is the set of necessary (and jointly sufficient) conditions for a generalisation to express a law.

[2]See also Lewis (1999), pp. 41–43 and pp. 233–244.

[3]See also Kistler (2002).

argument for the (conditional) necessity of laws (i.e., the thesis that a property cannot exist in a possible world without all the laws that hold of that property in the actual world) rather yield the conclusion that there are no laws at all.[4]

In this chapter, I defend the idea that laws have the same metaphysical status as natural properties. Science requires postulating both natural properties and laws. I take it that the best justification for metaphysical theses is that their truth is the best way to make sense of science. Quine's criterion for existence, in terms of being the value of a quantified variable in some scientific statement,[5] is too narrow. The mere fact that a variable figures in a well-confirmed law, I contend, warrants the metaphysical claim that the variable corresponds to a real property. Contrary to Quine's criterion, it is not necessary that there be scientific statements that quantify over the variable. Why do we believe that there is a property of having mass? The following justification seems sufficient. Mass is expressed by a variable that figures in well-confirmed laws such as Newton's law of universal gravitation $F = {}^{gM \cdot M^*}\!/_{r^2}$. This reasoning applies to properties and laws at all levels and in all sciences. In biophysics, e.g., it has been discovered that muscle fibres exhibit voltage oscillations that can be mathematically modelled by differential equations (Morris and Lecar 1981). To the extent that these models correctly approximate the behaviour of real muscle fibres, e.g. the giant muscle fibres of the barnacle (a marine arthropod), we may suppose that there are laws relating the properties that are represented by the variables featuring in the equations of the model. Thus, the fact that these equations contain (among many other variables) a variable that represents the conductance g_K of the muscle fibre's membrane for potassium ions gives us a reason for taking the property of having conductance g_K to be a real property of such membranes.

Being metaphysical theses, both the existence of properties and the existence of laws are controversial. My aim in this chapter is to defend the existence of laws against recent arguments that the existence of properties is more fundamental, so that laws have either only a derivative status or do not exist at all in addition to properties. It has been claimed that within the framework of a dispositionalist account of natural properties, according to which natural properties are powers or dispositions (or collections of such powers or dispositions), it is not necessary to postulate laws, because properties conceived as powers do all the explanatory work that is usually attributed to laws. I will argue against this thesis and show that the metaphysical status of laws is the same as that of natural properties, even if the latter are conceived in terms of powers. My main thesis is that our belief in the existence of laws, such as Newton's law of gravitation or the laws governing voltage oscillations in muscle fibres, is warranted by a reasoning of the same sort and of the same strength as our belief in the existence of natural properties, such as the property of

[4]Bird (2007) argues that they lead to a weaker conclusion: Laws supervene on natural properties, in such a way that knowledge of laws can be derived from knowledge of properties.

[5]"We may be said to countenance such and such an entity if and only if we regard the range of our variables as including such an entity. To *be* is to be a value of a variable" (Quine 1939/1976, p. 199; italics Quine's).

having mass (of a certain quantity) M or the property of cell membranes of having conductance g_K for potassium ions.

11.2 The Distinction Between Powers and Dispositions

My argument for the thesis that laws are required to make sense of science, even within a metaphysical framework that admits the existence of natural properties conceived as powers, has two steps.

I argue first that the truth-maker of disposition attributions contains natural properties, which I suggest conceiving as powers. This argument relies (1) on the thesis that the distinction between the dispositional and the categorical is a distinction between sorts of concepts or predicates, rather than between sorts of properties, and (2) on the thesis that natural[6] properties, conceived as powers, are distinct from dispositions[7]: for typical scientific properties, the same power figures in the truth-maker of many disposition attributions. In a second step, I argue that laws are required over and above powers, as constituents of the truth-makers of disposition attributions.

I shall take it for granted that (1) there is a conceptual link between the attribution of a disposition and the truth of a counterfactual conditional, but (2) the truth condition of a disposition attribution can only be adequately expressed by a counterfactual containing a ceteris paribus clause such as "in otherwise normal circumstances".[8] Object x has disposition D if and only if, "if *x* were in (triggering) circumstances T, then ceteris paribus, *x* would manifest M". Attributing to an object *x* the common sense disposition of being fragile means that, if *x* were sharply struck or subjected to appropriate stress, then, in otherwise normal circumstances, it would break. In the same way, attributing to a massive object *x* the disposition to fall with constant acceleration near the surface of the Earth means that, if *x* were dropped above but near the surface of the Earth, then it would, if no forces other than gravitational attraction acted on it, fall with constant acceleration in the direction of the centre of the Earth.

I will also take it for granted that the truth of the attribution of a disposition D to an object *x* requires that the object has an intrinsic property B, which is the causal basis of the disposition (Prior et al. 1982; Lewis 1997). It is true to say that *x* has the disposition to M if triggered by T only if *x* has some intrinsic property B (or set of intrinsic properties) that contributes causally, together with T, to bring about

[6]I will henceforth drop the qualifier "natural". It should be understood that what I say does not apply to "abundant" (Lewis 1983) properties, in the sense in which there is an abundant property for any arbitrary predicate, even for negative, disjunctive or "gerrymandered" predicates, such as "grue".

[7]Bird (Chap. 10, this volume) provides independent reasons for distinguishing between dispositions and powers.

[8]See Lipton (1999), Martin (1994), Bird (1998).

M. Then B is the causal basis, in the object x, of the disposition D to manifest M in conditions T. I suggest calling this causal basis the "power" underlying the disposition. Thus, the truth-maker of the attribution of the disposition D to x at t contains the fact that x has B at t and for some sufficiently extended period after t. If no intrinsic property of x contributes to M, it is not correct to ascribe the disposition to x. It is, e.g., not the placebo pill that can properly be said to have the disposition to heal; rather, what has that disposition is the doctor's act of prescribing it. Consider what generates the force on an electrically charged object in an electric field: the triggering condition (i.e., the presence of the electric field E) is not in itself causally sufficient to bring about the force; it is the interaction of the property of being charged with the field that produces the force.

The dispositional-categorical distinction can either be interpreted as an ontological distinction between sorts of properties or as a semantic distinction between sorts of concepts or predicates. I will here accept the latter conception according to which it corresponds to a distinction between two ways of *conceiving* of a given property: A dispositional concept (or predicate[9]) is a concept such that its attribution to some object *a priori* entails a certain counterfactual conditional. "A priori" means that the entailment is guaranteed by the content of the concept (the meaning of the predicate), so that it is (at least implicitly) known by every competent speaker, without any need for empirical information. By contrast, a concept is categorical insofar as its attribution to an object does not a priori entail any counterfactual conditional.

The property of bearing elementary electric charge can be conceived in a categorical way as a fundamental natural property. It can also be conceived in a dispositional way. Attributing to an electron x the property of having elementary electric charge is equivalent to the counterfactual: If x were placed in an electric field of strength $\vec{E}$ (triggering condition T), x would be subject to a force $\vec{F} = q\vec{E}$ (manifestation M).

This gives us a simple reason for distinguishing between dispositions and powers. "Power" is an ontological category that picks out natural properties, whereas "disposition" is a semantic category that picks out a way of conceiving and describing natural properties. All powers, i.e., natural properties, can in principle be conceived both in a categorical way as intrinsic properties of objects, or in a dispositional way, by way of the manifestations they would bring about if an object having them were in a certain triggering condition.

The thesis that all natural properties can be conceived both in a categorical and in a dispositional way bears some resemblance to the view that properties have a "dual nature": "Every property [...] endows its possessor with both a particular disposition or 'causal power' and a particular quality" (Heil 1998, p. 181). Here, the distinction between "disposition" and "quality" is taken to be ontological. However, this "dual aspect" theory does not explain how the dispositional and the qualitative aspect can coexist within the same property although they seem to be incompatible. This

[9]In what follows, it should be understood that what I say about concepts is also supposed to apply to predicates.

problem is solved if the distinction is taken to be semantic. Dispositional and categorical predicates can make reference to the same property although they have, as predicates, incompatible properties: dispositional predicates a priori entail a ceteris paribus counterfactual, whereas categorical predicates do not.

Do we really need both concepts of power and disposition, instead of just one? In other words, couldn't we simply say that a disposition is a power, and that a power is its own causal basis? The reason for which the distinction between power and disposition is indispensable is that many powers are "multi-track", i.e., contribute to make true different disposition attributions.

Let us take electric charge. The fact that an object x possesses the elementary electric charge q contributes to making true many disposition attributions to that object. Here are four of them:

1. The disposition of x to undergo a force $\vec{F} = q\vec{E}$ if x is in electric field $\vec{E}$.
2. The disposition of x to attract a second object with charge q* and at distance r with force $F = k_e \frac{q \cdot q^*}{r^2}$ (Coulomb force).
3. The disposition of x to undergo a force $\vec{F} = q\vec{v} \times \vec{B}$ (Lorentz force) if x is moving with $\vec{v}$ in $\vec{B}$.
4. The disposition of x to bear a magnetic moment $\vec{\mu} = \frac{1}{2} q\vec{r} \times \vec{v}$ if x is in rotation with speed $\vec{v}$ on radius $\vec{r}$.

The four dispositions mentioned are not identical: it is not the same thing for x to be disposed to undergo a force $\vec{F} = q\vec{E}$ (manifestation M_1) when placed in an electric field $\vec{E}$ (triggering condition T_1), for x to be disposed, at a distance r from of a second object that is charged with charge q* (triggering condition T_2), to exert on this object a force $F = k_e \frac{q \cdot q^*}{r^2}$ (manifestation M_2), and for x to be disposed to undergo a force $\vec{F} = q\vec{v} \times \vec{B}$ (manifestation M_3) when x moves with $\vec{v}$ through a magnetic field $\vec{B}$ (triggering condition T_3).

The very fact that these dispositions are different may seem paradoxical because, although they are different, they are all dispositions of electric charge, and electric charge seems to be just one power, not many. As long as we have only one concept of disposition/power, charge seems to be both one and many dispositions/powers.

A way of avoiding this paradox is suggested by the fact that these dispositions always go together or, in other words, that it does not seem to be possible that something has just one (or just two) of these dispositions without also having all the others. It is a fact that everything that undergoes the Lorentz force when it moves through a magnetic field is also subject to $\vec{F} = q\vec{E}$ when it is in field $\vec{E}$. It is a fact that there is no property that gives its bearer only one (or two) of these dispositions without also giving it the others. This indicates that something unites the set of dispositions. What is the relation between the property of bearing electric charge and the different dispositions D_i it gives its bearers? Being charged cannot be identical to all dispositions D_i, because the D_i are not identical with each other. Being charged cannot be identical to one of the D_i (but not the others) either, because it is not linked more closely to one of the D_i than to the others.

The paradox that charge is both one and many dispositions/powers is avoided if there is an element of reality, the natural property of being charged, which is both distinct from each of the dispositions in the set and nevertheless explains the fact that they are always present together. Charge is a powerful property, or simply "a power", which is not identical to the dispositions in the set. What is the link between the power and the dispositions? Power is a metaphysical concept, whereas disposition is a semantic concept. What is it in reality that makes the attribution of the different dispositions D_i to a given object *x* true? My hypothesis is that what unites the set of dispositions corresponding to electrical charge is that the truth-makers of the D_i have a common element.

In general, if several dispositions D_i can truthfully be attributed to an object *x*, each D_i to manifest M_i will have its own causal basis B_i (i.e., its own set of intrinsic properties of *x* that contribute causally to M_i, along with triggering circumstances T_i and background conditions). The different dispositions linked to electric charge we have mentioned above always come together, i.e., the attribution of all the dispositions in that set (D_i) to a given object always has the same truth value – in the sense that if one is true, all the others are true and if one is false all the others are false. It is a remarkable fact that there is nothing that has only one (or two) of these dispositions but not the others. This fact can be explained by our hypothesis that (1) the truth-makers of these disposition attribution have the powerful property of being charged as a common part and (2) the remaining parts of their truth-makers are constituted by laws that come as a metaphysical package: if one exists in a given possible world, all the others exist there too.[10]

Object *x* has the power of being charged if and only if there is an intrinsic property of *x* that contributes causally to the manifestations of all the dispositions associated with being charged: attracting other charged bodies, interacting with a magnetic field if in motion, etc. This power is the common causal basis of all these dispositions. If single-track dispositions exist, they are a special case of multi-track dispositions. A single-track disposition would be a disposition that is not lawfully linked to other dispositions. This seems never to be the case for scientific dispositions: laws of nature link different natural properties.[11]

Powers are theoretical properties. This implies that the postulate of their existence is fallible, and is justified by the usual criteria of theory construction. The two main reasons for postulating a theoretical property are that its existence provides the best unifying explanation of phenomena and that it is fruitful in suggesting new hypotheses. The reason to postulate the existence of electrical charge q is that the hypothesis of the existence of this intrinsic property provides the best unifying explanation of why the different dispositions (1) to (4) mentioned above always come together.

It is important to distinguish the concepts of causal basis and reduction basis. Insofar as all dispositions have manifestations, all dispositions have causal bases, which are powers. However, the causal basis of *x*'s disposition to exhibit

[10]This thesis will be justified in Sect. 11.3.

[11]We shall come back later to the significance of this fact for the metaphysical status of laws.

manifestation M_i in triggering situation T_i need not be microscopic[12] relative to the level of description of T_i and M_i. The disposition of a macroscopic object *x* to exert, in the vicinity of a second macroscopic object *y* that is electrically charged, a force on *y*, has a macroscopic causal basis: the property of *x* to be electrically charged. True, the macroscopic property of being charged can be reductively explained in terms of microscopic properties of the microscopic parts of *x*. But this need not be the case. Fundamental dispositions have no reduction basis, but they have a causal basis (Mumford 2006; Molnar 1999, 2003, ch. 8; Kistler 2012). The causal basis of the disposition of an electron to repel a second electron is the first electron's intrinsic property of bearing the elementary electric charge. Being elementary, it cannot be reduced to properties of the electron's parts, for there are no such parts. This shows that the causal basis of a disposition need not be micro-reducible. Generalizing to include macroscopic dispositions and their macroscopic causal bases, we get: the micro-reducibility of a macroscopic causal basis is not required for the truth of the disposition attribution. The reality and causal efficacy of the causal basis is independent of its micro-reducibility.

Many authors assume that multi-track powers can always be analysed in terms of single-track powers, so that only the latter should play a fundamental role in the metaphysical theory of powers and dispositions. Bird claims that "we do not need to posit fundamental multi-track dispositions" (Bird 2007, p. 24). His argument proceeds through two theses I will mention in a moment. What Bird calls "pure" and "impure" dispositions are equivalent to our "single-track" and "multi-track" powers. Attributing to object *x* the *pure* disposition D is equivalent to a single counterfactual conditional.

$(Dx \text{ and } Tx) \mathrel{\Box\!\!\rightarrow} Mx$.

If *x*, which has disposition D, were in triggering situation T*x*, it would manifest M*x*.[13] By contrast, attributing to object *x* the *impure* disposition I is equivalent to a whole set of conditionals:

For all i $[(Ix \text{ and } T_ix) \mathrel{\Box\!\!\rightarrow} M_ix]$.

In this vocabulary, Bird's first thesis is that.

(T1) All impure dispositions are conjunctions of pure dispositions.

In our vocabulary, (T1) says that it is equivalent to attribute a multi-track disposition to an object and to attribute a conjunction of single-track dispositions to it. Here are two reasons for thinking, against (T1), that the attribution of a multi-track power such as electrical charge has a content that goes beyond the attribution of

[12]Many authors take it for granted that the causal basis of the attribution of a macroscopic disposition to a macroscopic object must be microscopic, or "microstructural" (Quine 1971; Armstrong 1973; Prior et al. 1982).

[13]We shall here follow Bird in omitting the ceteris paribus qualifier, which must strictly speaking be inserted into the consequent of the counterfactual, in order to take account of the possibility of antidotes.

a determinate set of dispositions, each of which is defined by a (ceteris paribus) counterfactual conditional linking one triggering condition to one manifestation:

1. Multi-track powers are theoretical properties that provide a unifying explanation of a set of dispositions. If the multi-track (or impure) disposition were just equivalent to a certain conjunction of dispositions in a set (D_i), the fact that the dispositions belonging to (D_i) always come together would be a brute fact. However, if a multi-track power is a common part of the truth-makers of all the dispositions in (D_i), and if the laws constituting the remainder of the truth-maker "come in a package", this explains why these dispositions always come together.
2. The conception according to which a multi-track power is a common part of the truth-makers of the dispositions in a set (D_i), rather than being equivalent to the conjunction of these dispositions, allows making sense of scientific discoveries. It may be discovered after the multi-track power has been introduced (on the basis of set (D_i)) that there is a larger set (D_i^*) that contains (D_i) as a subset, such that all dispositions in (D_i^*) always come together. This could have happened if electrical charge was first introduced to provide a unifying explanation for why dispositions (1), (2) and (3) always come together, and if subsequently it was discovered that there is a larger set (D_i^*) that contains, over and above the dispositions (1), (2) and (3), a fourth disposition (4) to give rise to a magnetic moment, that is also always associated to each of the dispositions in set (D_i). This possibility shows that the power is not equivalent to any finite set of dispositions, such as (D_i) or (D_i^*). It is characteristic of theoretical properties that it remains always possible that new observable consequences are discovered.

Bird's second reason for not taking impure dispositions metaphysically seriously is that (T2) "All impure dispositions are non-fundamental" (Bird 2007, p. 22). If it were correct that all impure (or multi-track) dispositions are non-fundamental, they would be micro-reducible in terms of fundamental dispositions, which are pure (or single-track). However, as I have argued above, the issue of micro-reducibility is independent of the issue of the possibility of a unifying explanation of a set of dispositions.

11.3 Laws as Parts of the Truth-Makers of Disposition Attributions

We are now in a position to draw the implications of the conceptions of powers and dispositions developed so far on the issue of the metaphysical status of laws. Laws are indispensable to make sense of what science tells us about dispositions, for two reasons:

The first reason is that laws are required to explain the structure of the dispositions corresponding to one property, or power. The second reason is that dispositional concepts and predicates are relational, whereas powers are typically monadic. The truth-maker of a disposition attribution must contain a relational part over and above powerful properties, which are monadic. Laws provide this relational part.

I will argue for the existence of laws alongside powerful properties, by showing that the rival thesis, according to which laws do not exist (Mumford) or can be derived from powers (Bird), leads to inacceptable consequences. Bird (2007) argues that the metaphysical status of laws is derivative relative to natural properties, on the basis of the following three theses: First, natural properties are powers, in the sense that the possession of such a property necessarily entails the possession of a disposition. Second, at least some laws are metaphysically necessary,[14] in the sense of holding in all possible worlds in which the properties figuring in them exist. Third, these laws are entailed by the properties. From these premises, Bird draws the conclusion that a universal generalisation holding for objects possessing a property P, which is equivalent to a law, can be "derived from a claim about the essence of P" (Bird 2007, p. 46). He concludes that laws have only a derivative status with respect to properties. Once powerful properties have been postulated, there is no need to postulate laws in addition.[15]

Mumford (2004) argues for a "lawless" metaphysics. From a metaphysical point of view, there are no laws. The role that has been attributed to laws by nomological realists, such as making true scientific explanations and predictions and justifying counterfactuals, is played by natural properties.[16] The natural properties postulated by scientific hypotheses and theories can play this role because they are "modally involved" in the sense that they contain necessary relations to other properties. Thus, at every occasion at which such a property is instantiated, the object possessing the property stands in a necessary relation to other properties. From these premises, Mumford draws the conclusion that it is a mistake to postulate laws over and above natural properties, for properties alone, conceived as powers, can do all the metaphysical work needed: explain what makes true scientific explanations and predictions and what justifies counterfactuals.

[14]"Some laws must be held to be necessary whether one is a categoricalist or a dispositional essentialist" (Bird 2007, p. 177).

[15]Physicists do certainly not derive laws from powers. Instead, contemporary physics uses symmetries (Livanios 2010, 2018a; French 2014, Chap. 12, This volume) and extremal principles such as Hamilton's principle (Livanios 2018b). This observation fits well with my thesis that laws are needed to make sense of science. My aim in this paper is to argue that powers alone are not sufficient for making sense of science. However, I will also offer reasons for the thesis that laws alone (or the symmetries and other principles from which the laws are derived) are not sufficient by themselves for making sense of science.

[16]True, this is not how Mumford himself would present his thesis. He claims: "I am not making properties serve the role of laws" because "the world leaves no such role to be filled by anything" (Mumford 2004, p. 204). However, this is intended to be true of laws conceived as "governing", from the outside as it were, properties that would otherwise, if it were not for the laws, be inert. If I nevertheless attribute to Mumford the thesis that powerful properties fill the role that has traditionally been attributed to laws, I mean the role of justifying counterfactuals and to make possible scientific explanation and prediction of phenomena.

My argument against such a lawless metaphysics, and against Bird's thesis that laws are implicit in powers and can be derived from them, rests on the claim that laws are required to account for the relation between properties and dispositions. Mumford considers two ways of conceiving the relation between a property P and the dispositions P gives an object:

1. According to the first conception, properties are sets of dispositions. In other words, dispositions are *elements* of properties. To evaluate this thesis, it will be useful to translate Mumford's vocabulary into the vocabulary used in this chapter. What Mumford calls "properties" are our powers (or powerful properties): both express the metaphysical concept of what it is about an object that makes true the attribution of a disposition or a counterfactual linking a triggering condition to a manifestation condition. Mumford's powers are our dispositions: each power (disposition) corresponds to exactly one counterfactual linking a triggering condition to a manifestation. In our vocabulary, the thesis that "properties are sets of instantiated powers" (Mumford 2004, p. 171) translates as the thesis that "powerful properties are sets of dispositions".
2. Mumford considers a second way of conceiving the relation between a property P and the dispositions D_i. According to this conception, properties are "mereological sums of" (Mumford 2004, p. 171) powers. In our vocabulary, powerful properties are mereological sums of dispositions.

If one of these conceptions is correct, it follows indeed that no law is required for making true the attribution of a disposition. The mere possession by an object x of P is sufficient to make true the attribution to x of every disposition that is an element (conception 1) or a part (conception 2) of P; the mere possession of P by x is also sufficient to make true, for each D_i in P, its characteristic counterfactual "if x were in condition T_i, it would, ceteris paribus, manifest M_i".

I will call the conception of properties according to which a property *contains* all the dispositions (either as elements or as parts) it gives every object possessing it, the *thick conception of properties*.

According to this thick conception of properties,

1. a property P *contains* all/some of the dispositions D_i that P gives its bearers. An object x's property of being electrically charged, e.g., contains x's disposition to undergo a force $\vec{F} = q\vec{E}$ if x is in electric field $\vec{E}$.
2. The electric field $\vec{E}$ and the force $\vec{F}$ are in some sense "contained" in P.
3. If we accept the view that properties are internally related (I will argue in a moment that laws are such internal relations), then P also contains these internal relations to other properties such as $\vec{E}$ and $\vec{F}$.

We would have a strong argument for the thick conception of properties if it followed from the thesis that laws of nature are conditionally necessary. Let me sketch the reasoning that leads to adopting this thesis. It will turn out that it does not favour the thick conception over the alternative "thin" conception, according to which natural properties contain neither the laws featuring them nor the dispositions they give objects having them. The thesis that laws of nature are conditionally necessary is a consequence of the following reasoning about what determines the identity of a natural property. Nothing else than the laws a property figures in can determine the identity of a property. What is it to be electrically charged? Insofar as being electrically charged is a property that has been introduced on theoretical grounds we know exactly as much about this property as we know laws featuring it. Scientific hypotheses being fallible, the laws that we take to be true may turn out to be incorrect; and there are certainly laws featuring electrical charge that have not yet been discovered. In other words, the laws we take to be true of charge today are an approximation of a subset of the laws that really hold of electrical charge. However, our partial ignorance notwithstanding, insofar as the law of the Coulomb force is approximately true, this law expresses part of what it is to be electrically charged: it is to attract (or repel) other objects at distance r that are charged with q* with force $F = k_e \frac{q \cdot q^*}{r^2}$. In the same sense, part of what it is to be electrically charged with charge q is to undergo a force $\vec{F} = q\vec{E}$ if placed in an electric field $\vec{E}$. The same holds for the other laws featuring electrical charge.

If there is a possible world in which there is a quantitative property q+ such that objects possessing q+ attract or repel each other with a force proportional to the product of their quantities of q+ and to the inverse *cube* of their distance, then q+ is *not* the property of being electrically charged. Similarly for all other natural properties: it is part of what it is to be massive that a massive object attracts other massive objects according to the law of universal gravitation $F = {}^{gM \cdot M^*}\!/_{r^2}$. Thus, if some possible world contains a property M+ such that two objects having respectively quantities M_1+ and M_2+ of this property M+ *repel* each other with a force proportional to (the product of their quantities of M+ and) the inverse of the square of their distance, then M+ is *not* the property of having mass M. The same reasoning applies to all properties M++ featuring in laws that do not have *exactly* the same form as the law of universal gravitation $F = {}^{gM \cdot M^*}\!/_{r^2}$. In other words, in all possible worlds in which the property of having mass M exists, it is such that pairs of objects having M *attract* each other with a force proportional to (the product of their quantities of M and) the inverse of the square of their distance. It is important to stress that this argument only leads to the consequence that laws are *conditionally* necessary: for all possible worlds w, *if* the property of having mass M exists in world w, then all objects possessing this property obey to the inverse square law of gravitational attraction. The law is not absolutely necessary because there may well be worlds in which there is no such property at all.[17]

It follows from this conception of laws as conditionally necessary that every world containing P contains all laws featuring P that exist in the actual world w_0; furthermore, in all these worlds, all properties lawfully linked to P also exist. In other

[17]In a world in which M does not exist, all universal conditional statements whose antecedent attributes M are trivially true. From an empiricist point of view, in which laws are statements, one might want to say that laws bearing on M are trivially true in all worlds in which there is no property

words, laws and natural properties come in a package: they form a network that is characteristic of the actual world w_0; for all possible worlds w, either w contains both *all* the natural properties of the actual world w_0 and *all* the laws holding in w_0, or *none*.

It also follows that laws are internal relations between properties. An internal relation R between b and c is such that b and c could not exist without being related by R(b,c). In other words, in every possible world in which b and c exist, they are related by R. Take two properties P and Q that exist at world w. If there is a law L relating P and Q in w, then this law exists in every world in which P and Q exist. This implies that P and Q are internally related by L.

However, it does not follow that any property $Q \neq P$, even if is lawfully linked to P, is *part* of property P. Even if Q is necessarily linked to P (because Q exists in every world in which P exists, because every world in which P exists contains a law linking it to Q), it does not follow that Q is *part of* P or somehow *contained in* P.

We are now ready for four arguments for the thesis that properties are not just sets, clusters or mereological sums of dispositions, and for postulating the existence of laws. Here is the first. Laws explain the structure of the set of dispositions whose attribution is made true by property P. Such a structure cannot be explained in the framework of a conception according to which properties are just sets (or mereological sums) of dispositions.

Recall that the possession of a natural property such as having electrical charge q makes true a whole set of disposition attributions:

1. If an object with q were in $\vec{E}$, it would undergo $\vec{F} = q\vec{E}$.
2. If an object with q were moving with $\vec{v}$ in a magnetic field $\vec{B}$, it would undergo a force $\vec{F} = q\vec{v} \times \vec{B}$.
3. If an object with q were in rotation with speed $\vec{v}$ on radius $\vec{r}$, it would create magnetic moment $\vec{\mu} = {}^{1}\!/\!_{2} q\vec{r} \times \vec{v}$ ....

These dispositions, which correspond to the laws in which q explicitly figures, are only the tip of an iceberg of dispositions whose attribution is made true by the possession of the property of being charged with q. The reason is that there are many dispositions whose attribution to an object x is made true by the conjunction of *several* laws: one law in which q itself figures, together with other properties such as R, and one or several other laws featuring R.

As an example, the attribution to electrons of the disposition to create a magnetic field if they are in motion as electric current I (Ampere's law) is directly made true by the property of the electrons of being electrically charged. The truth-maker of the attribution of this disposition has two parts: (1) the possession of charge by the electrons and (2) Ampere's law that links the movement of electric charge (electric current) to the existence of a magnetic field.

of having mass M. However, from a realist point of view, a law is a relation between properties. If one of the properties does not exist in a given world, the law itself does not exist there either.

But the same property of having electric charge also contributes to make true, in a more indirect way, the attribution to electrons (moving as electric current) of the disposition to exercise a force on a second wire carrying electric current. This disposition has a more complex truth-maker, which is composed of (1) the property of the electrons of having electric charge and two laws: (2) Ampere's law linking the current I_1 in the first wire to $\vec{B}$, and (3) the law of the Laplace force $\vec{F} = \vec{I}\,l \times \vec{B}$, linking $\vec{B}$, together with a current I_2 in a second wire, to a force $\vec{F}$ (with $l =$ the length of the second wire situated in field $\vec{B}$).

By virtue of the laws featuring $\vec{B}$, such as the law of the Lorentz force, electrons have more dispositions than just those corresponding to the laws directly featuring the property of being charged. Electrons have the disposition to create $\vec{B}$, which itself has the disposition to create a force on a second wire carrying electric current. So, indirectly, moving electrons have the disposition to create a force on other charges.

So here is my first argument against the thesis that P is a cluster of dispositions. A cluster is *unstructured*. Therefore, it cannot account for the fact that some dispositions are more directly linked to a property than others. In our example, the property of bearing electric charge gives a charged object *directly* the disposition to attract other charges, but only *indirectly* the disposition to exercise a force on distant wires carrying electric current.

Laws explain the difference between dispositions that are directly made true by a powerful property P and dispositions that are only indirectly made true by P. Disposition attributions are directly made true by P if they are determined by (1) power P and (2) a law that explicitly contains P. This is the case of the disposition of electric current to create magnetic field $\vec{B}$ according to Ampere's law. The set of laws directly containing property P makes true a first set S_1 of dispositions. This set can be compared to a "first shell of an onion centred on P". The dispositions in a second set S_2 are more indirectly made true by property P. This "second onion shell" contains dispositions whose attribution to an object x is made true by P and *two* laws: a law L_1 directly containing P and a law L_2 containing another property contained in L_1.

In our example, the powerful property of carrying electric current is lawfully linked to magnetic field $\vec{B}$ (Ampere's law plays here the role of L_1), whereas $\vec{B}$ is linked to force $\vec{F} = \vec{I}\,l \times \vec{B}$ by the law of the Laplace force. Thus the second set S_2 contains a disposition whose attribution to a wire w is made true by (1) the fact that electric current flows through w, (2) Ampere's law linking current to field $\vec{B}$, and (3) the law of the Laplace force linking $\vec{B}$ to a force on a second wire carrying electric current.

The same reasoning shows that there are many more sets of dispositions S_3, S_4 etc. linked even more indirectly to the possession a given powerful property, by three or more laws in addition to the possession of that property.

Here is a second argument against the thesis that a powerful property P is just a set of dispositions. This doctrine cannot explain why there are no properties corresponding to *subsets* of these sets. Why is there a cluster containing D_1 "if $\vec{E}$ then $\vec{F} = q\vec{E}$", D_2 "if $\vec{v}$ and $\vec{B}$, then $\vec{F} = q\vec{v} \times \vec{B}$", and D_3 "if $\vec{v}$ around circle $\vec{r}$, then $\vec{\mu} = \frac{1}{2}q\vec{r} \times \vec{v}$", but no cluster containing, say, only the first two of these three dispositions? Why isn't there a property Q that gives its bearer the dispositions to undergo $\vec{F} = q\vec{E}$ if in electric field $\vec{E}$ and the disposition to undergo $\vec{F} = q\vec{v} \times \vec{B}$ if moving with $\vec{v}$ through field $\vec{B}$, but that does *not* give its bearer the disposition to produce magnetic moment $\vec{\mu} = \frac{1}{2}q\vec{r} \times \vec{v}$ if rotating with speed $\vec{v}$ around radius $\vec{r}$? Postulating laws makes it possible to answer this question: the set of laws containing the property P exactly determines the set of dispositions whose attribution is made true by P. A property such as Q cannot exist: an object can have D_1 only if it has q and if the law $\vec{F} = q\vec{E}$ holds. But, if laws are conditionally necessary, every world in which q exists contains *all* the laws containing q. It contains in particular the law by which circulating charges create magnetic moment $\vec{\mu} = \frac{1}{2}q\vec{r} \times \vec{v}$. So every object that has disposition D_1 because it has q (and because the law $\vec{F} = q\vec{E}$ holds) must also have D_3 because it has q and because the law of magnetic moment holds in all worlds containing q.

At one point, Mumford seems to acknowledge that a property isn't just a set or a cluster because there are "real and important connections" (Mumford 2004, p. 173) between the dispositions in a cluster. However, he argues that these connections do not correspond to laws, because laws would have to be external relations. This argument has no force against our view that laws are internal relations between properties.

A third argument against taking properties P to be clusters of dispositions (and against denying the role of laws in determining dispositions) is that it does not seem to be possible, without laws, to make sense of systematic relations between different properties/clusters. Here is why.

Suppose, for the sake of the argument, that properties are clusters of dispositions. Then it is systematically the case that the same disposition belongs to several clusters. Both the cluster of $\vec{v}$ (speed) and the cluster of $\vec{B}$ (magnetic field) contain, e.g., the disposition to produce a force on moving charged objects proportional to their charge and speed. In the cluster conception, these dispositions are parts of different properties, $\vec{v}$ and $\vec{B}$.

If properties were clusters, one such cluster, say the cluster of $\vec{v}$, should put no constraints on other clusters, such as the cluster of $\vec{B}$. However, dispositions belonging to different clusters constrain each other. Given that $\vec{v}$ gives a charged object the disposition to undergo $\vec{F} = q\vec{v} \times \vec{B}$ if in $\vec{B}$, the disposition of $\vec{B}$ to exert a force on charges moving with speed $\vec{v}$ must be the disposition to produce $\vec{F} = q\vec{v} \times \vec{B}$. $\vec{B}$ could not contain, say, the disposition to produce $\vec{F} = q^3\vec{v} \times \vec{B}$.

If we take properties to be thin, and to be related to other properties by laws, we can explain these constraints as nomological. Dispositions figuring in different clusters can be constrained by each other because they are determined by the same law. The attribution to an object moving with speed $\vec{v}$ of the disposition to undergo

Lorentz force ($\vec{F} = q\vec{v} \times \vec{B}$) is made true by the object's property of having speed $\vec{v}$ and by the law of Lorentz force. The attribution to field $\vec{B}$ of the disposition to exert force $\vec{F} = q\vec{v} \times \vec{B}$ on particles with charge q and speed $\vec{v}$ is made true by $\vec{B}$ and *the same law* of Lorentz force. The fact that the same law contributes to make true both disposition attributions explains why these dispositions are not independent of each other.

Here is a fourth and last argument for taking laws to be part of the truth-maker of disposition attributions (over and above powers). The truth-maker must be in part relational because the disposition is relational, whereas the relevant powerful property is monadic. At least some dispositions have relational triggering and manifestation conditions. In the case of the property of being charged, the triggering condition of the disposition, attributed to an object x bearing charge q, to attract other objects, is a relational fact about x: the fact that x is at distance r from a second object y with electric charge q*. The manifestation condition is also a relational fact: that x exercises a Coulomb force $F = k_e \frac{q \cdot q^*}{r^2}$ on y. The truth-maker of the attribution of this disposition to x cannot be entirely monadic because the triggering and manifestation conditions of the disposition are relational. Part of the truth-maker is the powerful property of bearing electrical charge, which is monadic. If there are laws, we can conceive of the truth-maker as composite, having a monadic part, which is the powerful property P and a relational part, which is the law relating the power P and the triggering condition to the manifestation condition.

11.4 Conclusion

I have sketched a metaphysical framework in which powers and laws together play the metaphysical role of making true disposition attributions. This framework explains why dispositions are systematically correlated with each other. Natural properties that feature in several laws are a common truth-maker for dispositions that always go together. Such properties are powers that correspond to the traditional concept of a multi-track disposition, which we have reinterpreted as consisting of a set of disposition attributions whose truth-makers have a common part. The common part of their truth-makers is a power that is the causal basis of each of the dispositions: it is an intrinsic property of the object to which the disposition is truthfully attributed, which causally contributes, together with a triggering condition, to bringing about the corresponding manifestation.

The reasons for postulating such powerful properties are the same as the reasons for postulating theoretical properties in general: such a postulate is justified insofar as it is the best unifying explanation of the fact that certain disposition attributions are always true together, and insofar as it is fruitful in suggesting yet unknown dispositions.

Disposition attributions have composite truth-makers. What makes it true that object x has the disposition to M in triggering situation T, is (1) a monadic powerful property P possessed by *x*, which is the causal basis of the disposition, and (2) one or several laws relating T and P to M. Powers are natural properties which are represented by the variables figuring in scientific laws. Against the thesis that laws are not needed over and above powerful properties to make sense of the scientific explanation, we have argued for the following theses. Laws are internal relations between properties. Natural properties make true disposition attributions only together with laws. Natural properties neither contain (as parts or as elements) the laws that link them to other properties nor those other properties. The postulate of laws over and above natural properties is required to explain the following features of the dispositions grounded on each natural property P. (1) The set of dispositions grounded on P is structured into dispositions that are more or less directly linked to P. (2) If a property P grounds the set D of dispositions, there are no properties that ground proper subsets of D. (3) Different properties often constrain each other's identity. (4) The truth-maker of many disposition attributions is relational even if the disposition is grounded on a monadic property.

Acknowledgements I thank my audiences in Innsbruck, Geneva and Louvain-la-Neuve, where I have presented earlier versions of this chapter, and an anonymous reviewer for this volume, for their helpful comments. Research for this chapter has benefited from financial support by ANR (grant ANR-12-BSH3-0009).

References

Armstrong, David M. 1973. Beliefs as States. In *Dispositions*, ed. R. Tuomela, 411–425. Dordrecht: Reidel.

———. 1983. *What is a Law of Nature?* Cambridge: Cambridge University Press.

Bird, Alexander. 1998. Dispositions and Antidotes. *The Philosophical Quarterly* 48: 227–234.

———. 2006. Looking for Laws. Symposium Review on S. Mumford, *Laws in Nature*, Abingdon: Routledge, 2004, by B. Ellis, A. Bird, and S. Psillos with a Reply by S. Mumford. *Metascience* 15: 441–451.

———. 2007. *Nature's Metaphysics. Laws and Properties*. Oxford: Oxford University Press.

———. This volume. A Dispositionalist Account of Causation, with Some Remarks on the Ontology of Dispositions. In *Dispositionalism. Perspectives from Metaphysics and the Philosophy of Science (Synthese Library* 417*)*, ed. A. S. Meincke, 151–170. Cham: Springer.

Dretske, Fred. 1977. Laws of Nature. *Philosophy of Science* 44: 248–268.

French, Steven. 2014. *The Structure of the World*. Oxford: Oxford University Press.

———. This volume. Doing Away with Dispositions: Powers in the Context of Modern Physics. In *Dispositionalism. Perspectives from Metaphysics and the Philosophy of Science (Synthese Library* 417*)*, ed. A. S. Meincke, 189–212. Cham: Springer.

Goodman, N. 1955. *Fact, Fiction, and Forecast*. Indianapolis: Bobbs-Merrill.

Heil, John. 1998. *Philosophy of Mind*. London: Routledge.

Kistler, Max. 2002. The Causal Criterion of Reality and the Necessity of Laws of Nature. *Metaphysica* 3: 57–86.

———. 2012. Powerful Properties and the Causal Basis of Dispositions. In *Properties, Powers and Structures. Issues in the Metaphysics of Realism*, ed. A. Bird, B. Ellis, and H. Sankey, 119–137. New York/Oxford: Routledge.

Lewis, David K. 1973. *Counterfactuals*. Oxford: Blackwell.

———. 1983. New Work for a Theory of Universals. *Australasian Journal of Philosophy* 61: 343–377. Reprinted in Lewis 1999, 8–55.
———. 1997. Finkish Dispositions. *Philosophical Quarterly* 47: 143–158. Reprinted in Lewis 1999, 133–151.
———. 1999. *Papers in Metaphysics and Epistemology*. Cambridge: Cambridge University Press.
Lipton, Peter. 1999. All Else Being Equal. *Philosophy* 74: 155–168.
Livanios, Vassilis. 2010. Symmetries, Dispositions and Essences. *Philosophical Studies* 148: 295–305.
———. 2018a. Dispositionality and Symmetry Structure. *Metaphysica* 19: 201–217.
———. 2018b. Hamilton's Principle and Dispositional Essentialism: Friends or Foes. *Journal for General Philosophy of Science* 49: 59–71.
Martin, Charles B. 1994. Dispositions and Conditionals. *The Philosophical Quarterly* 44: 1–8.
Molnar, George. 1999. Are Dispositions Reducible? *The Philosophical Quarterly* 49: 8–17.
———. 2003. In *Powers: A Study in Metaphysics*, ed. S. Mumford. Oxford: Oxford University Press.
Morris, Catherine E., and Harold Lecar. 1981. Voltage Oscillations in the Barnacle Giant Muscle Fiber. *Biophysical Journal* 35: 193–213.
Mumford, Stephen. 2004. *Laws in Nature*. London: Routledge.
———. 2006. The Ungrounded Argument. *Synthese* 149: 471–489.
Prior, Elizabeth W., Robert Pargetter, and Frank Jackson. 1982. Three Theses About Dispositions. *American Philosophical Quarterly* 19: 251–257.
Quine, William V. O. 1939/1976. A Logistical Approach to the Ontological Problem. *Journal of Unified Science* 9: 84–89. Reprinted in his *The Ways of Paradox and Other Essays*, rev. ed., Cambridge, MA: Harvard University Press, 197–202.
———. 1971. *The Roots of Reference*. La Salle: Open Court.
Shoemaker, Sydney. 1984. Causality and Properties. In *Identity, Cause and Mind*, ed. S. Shoemaker, 206–233. Cambridge: Cambridge University Press. Reprinted in *Properties*, ed. D. H. Mellor and A. Oliver, 228–254. Oxford: Oxford University Press, 1997.
———. 1998. Causal and Metaphysical Necessity. *Pacific Philosophical Quarterly* 79: 59–77.
Tooley, Michael. 1977. The Nature of Laws. *Canadian Journal of Philosophy* 7: 667–698.

Chapter 12
Doing Away with Dispositions: Powers in the Context of Modern Physics

Steven French

Abstract Recent accounts of dispositionalism have extended this stance from vases to quarks. Here I shall present an obstacle to such an extension in the form of the symmetry principles that play such a fundamental role in the Standard Model of elementary particle physics. After considering certain ways the dispositionalist might get around this obstacle I shall argue that in its standard 'Stimulus and Manifestation' form, this stance should be abandoned. However, this does not mean one should give up on modal metaphysics entirely and adopt some form of Humeanism. Instead I shall suggest that current metaphysics presents certain 'tools' that might be deployed in this context, such as Vetter's recently developed notion of 'potentiality'. Thus there is still hope for a form of naturalised metaphysics that draws on extant metaphysical devices.

Keywords Dispositions · Potentiality · Spin · Standard Model · Symmetry

12.1 Introduction

The metaphysics of science has become something of a 'hot' topic in recent years with debates over the tri-partite relationship between metaphysics, science and the philosophy of science engendering a range of views. At one end of the spectrum we find a form of extreme naturalism that insists that metaphysics must cut its cloth entirely to fit the contours of the relevant science, typically fundamental physics. One can identify such a view in the opening chapter of Ladyman and Ross's *Every Thing Must Go* (2007), where they run through a variety of claims and positions in current metaphysics, picking them off one by one and dismissing them for failing to mesh with modern physics. A somewhat more moderate tone can be found in Callender (2011), but again the message is clear: philosophers of science have little to learn from metaphysicians who have signally failed to keep up with developments in science.

S. French (✉)
School of Philosophy, Religion and History of Science, University of Leeds, Leeds, UK
e-mail: S.R.D.French@leeds.ac.uk

A. S. Meincke (ed.), *Dispositionalism*, Synthese Library 417,
https://doi.org/10.1007/978-3-030-28722-1_12

This is not my stance. Together with Kerry McKenzie, I have tried to articulate what we have called the ‘toolbox’ approach to metaphysics, according to which metaphysics can be viewed as providing a set of tools that philosophers in other sub-disciplines, and particularly, the philosophy of science, can use for their own purposes (French and McKenzie 2012). This approach can be broadly divided into two phases: we begin by noting that modern metaphysics has, for whatever reasons, set off along a different path from much of the philosophy of science, which has taken it further and further away from what are generally understood to be the fundamental features of science, particularly physics. As a result, current discussions about gunk and simples, or even ‘core’ notions such as intrinsicality and fundamentality, seem to bear little relation to how scientists conceive of the world. However, we nevertheless think that metaphysics offers an array of moves and manoeuvres, devices and techniques, etc., that can be profitably deployed to elaborate an understanding of the world that does mesh with what science tells us, where we take that understanding to, minimally, go beyond a straightforward recitation of the scientific details.

I think this offers a useful way of tackling issues to do with the ontology of modern physics (French 2014). However, there is an obvious tension that arises, which can be summarised as follows: how can we distinguish between that metaphysics that can be profitably used as a ‘tool’ in understanding the world, and that which should be thrown away, given that we cannot foresee what new views of the world science, and in particular, fundamental physics, will come up with? The response is to note (McKenzie and French 2015), that the toolbox approach is conditionalised twice over: first, of course, upon naturalistically inclined metaphysicians taking tools from the toolbox, instead of making them ‘to order’ as it were, and thereby completely sidelining traditional, analytic, non-naturalised metaphysics; and secondly, upon those tools actually being useful to the interpretation of science as it develops.

Thus, the extent to which modern metaphysics can be viewed as relevant in the context of scientific developments is dependent upon features external to the enterprise itself – features that have to do with the evolution of both science and the philosophy of science. Nevertheless, whatever conditionalised support metaphysics gets from naturalistic projects, it remains the case that metaphysicians themselves should concede that the systematic disregard of real science that is so prevalent in current discussions simply cannot continue if they are to take their own projects seriously (ibid.).

With that in mind, here’s how the toolbox approach works in the context of this volume: within phase one, both dispositional and Humean accounts of laws and associated properties can be dismissed on the grounds that they fail to mesh with or even acknowledge in some cases certain features of fundamental physics. Here I shall focus on dispositionalism, as standardly understood (for criticism of Humeanism in this vein, see McKenzie 2013). In phase two, certain recent developments in the discussion over powers and dispositions may be examined with a view to their providing useful tools in constructing an appropriate account of modality in the context of modern science.

Thus the plan of the essay is as follows: in Sect. 12.2, I shall outline the standard dispositionalist account. I emphasise 'standard' because I recognise that, to put it flippantly, if you were to collect 10 defenders of dispositionalism or 'powers' views more generally and put them in a room, you would get 15 different accounts! Indeed, my aim is to draw on some features of certain non-standard perspectives later in the essay. I shall then note how this account has been extended from the everyday to the realm of modern physics – from vases to quarks, in effect. Here, however, there is a fundamental obstacle: the role of symmetries as constraints on the fundamental laws in physics, which I shall outline in Sect. 12.3. One of the great virtues of the standard dispositionalist account is that it supposedly yields laws from dispositions but it remains unclear, at best, how it can accommodate such symmetry principles.

I shall indicate some ways the dispositonalist might try to do that in Sects. 12.4 and 12.5 but in effect I will close off each option by indicating how the problem merely gets pushed back a step or two. Ultimately, the dispositionalist will have to ascribe her dispositions to objects that are metaphysically very 'thin', in the sense that they are mere placeholders within the physical structure, and also give up the standard Stimulus and Manifestation account, as I shall argue in Sects. 12.6 and 12.7. Of course, that will be music to the ears of certain other critics within and outwith the powers camp and I will conclude, in Sect. 12.8, by considering some recent proposals and evaluating whether they might be 'fit for purpose' within the context of modern physics.

12.2 Dispositionalism: The Standard Account

Let me begin with what I take to be the 'classic' statement of dispositionalism: Dispositional properties are those that play, as a matter of conceptual necessity, a certain causal role that is best captured in conditional terms (see, e.g., Mumford 1998). The equally classic example is that of fragility, understood in terms of the disposition to crack or break under appropriate circumstances, with the former taken to be the manifestation of the disposition and the latter circumstances, whatever they may be, as the stimulus. Thus the standard account is often framed in terms of the *Stimulus and Manifestation Characterisation (S&M)*: for a property to be dispositional is for it to relate a stimulus (S) and a manifestation (M) such that, if an object instantiates the dispositional property, it would yield the manifestation in response to the stimulus; or a little more formally: $\forall x((\mathrm{P}x \ \& \ \mathrm{S}\,x) \rightarrow \mathrm{M}x)$, where P is the relevant property, S the stimulus and M the manifestation, of course.

Here I shall set to one side all the much pored over issues about what counts as a stimulus, or a manifestation, the impact of finks, blockers and so on, simply because they are typically not relevant to the kinds of situations I shall be considering.

However, I shall mention, at least, what is taken to be a significant positive feature of this account, which is that it yields the laws of nature and thereby grounds their modal robustness (see Bird 2007). In particular, if one accepts what is sometimes called the Dispositional Identity Thesis, in terms of which the identity of properties is entirely cashed out in dispositional terms, so that, in effect, this identity is given by their place in the modal structure, then, adopting the above formal expression of S&M, laws can be seen to 'flow from' the relevant set of dispositions. And if we then consider a different possible world, populated by the same set of objects, then we will have the same properties, and hence by the Dispositional Identity Thesis, the same dispositions and the same laws, thereby accounting for the (physical) necessity of laws.

Over the last 20 years or so the account has been extended from everyday entities and their properties, such as vases and window panes, to the sorts of objects we find in physics, such as electrons and quarks. One of the most significant statements of such an extension, if not the earliest, can be found here:

> Physics tells us what result is apt to be produced by the having of gravitational pull or of electromagnetic charge. It does not tell us anything else about these properties. In the Standard Model the fundamental physical magnitudes are represented as ones whose whole nature is exhausted by their dispositionality: that is, only their dispositionality enters into their definition. Properties of elementary particles are not given to us in experience: they have no accessible qualitative aspect or feature. There is no 'impression corresponding to the idea' here. What these properties are is exhausted by what they have a potential for doing, both when they are doing it and when they are not. (Molnar 1999, p. 13)

There are of course many ways in which one can critique this statement. One can point out that it relies on a narrow understanding of 'experience' and argue that although quark confinement may prevent us from gaining epistemic access to single quarks, we can certainly 'experience' their properties, such as strangeness, beauty, charm etc., *via* quark jets and the like, where 'experience' is understood in non-empiricist terms. As for electrons and their property of charge, even 45 years ago (when I was studying physics at school), school-children had access to them (again, on a non-empiricist understanding of 'access') and their properties via reproductions of Millikan's famous oil-drop experiment! And of course it is an argumentative leap from a supposed lack of epistemic access to the denial of quiddities, which presumably are not intended to be accessible in the first place. More importantly, it is simply not the case that the nature of the fundamental properties of modern physics is exhausted by their dispositionality, at least not as the latter is understood on the standard account (see French 2013, 2014). Indeed, this is precisely the point I shall be exploring here.

Nevertheless, the extension of a dispositionalist stance to physics and the attendant claim that the fundamental properties of nature are (essentially) dispositional is now commonplace among dispositionalists of various stripes. Thus, Chakravartty asks,

> Why and how do particulars interact? It is in virtue of the fact that they have certain properties that they behave in the ways they do. Properties such as masses, charges, accelerations, volumes and temperatures, all confer on the objects that have them certain abilities or capacities. These capacities are dispositions to behave in certain ways when in the presence or absence of other particulars and their properties. (Chakravartty 2007, p. 41)

In similar vein, Mumford asserts that,

> [p]hysics in particular seems to invoke powers, forces and propensities, such as the spin, charge, mass and radioactive decay of subatomic particles (Mumford 2011, p. 267).

Here, just to continue with the somewhat churlish tone, one might wonder on what basis radioactive decay is set alongside spin, charge and mass, where the latter are fundamental properties, sometimes, and problematically (see French and McKenzie 2012) regarded as 'intrinsic', whereas the former is a phenomenon that results from quantum tunnelling. Nevertheless, that spin in particular is taken to be ultimately dispositional is something I shall be interrogating in some detail later on.

Before I get stuck into the physics it is worth noting that this broad picture of dispositions yielding laws has been attacked from a number of different directions. Thus Vetter, for example (Vetter 2009), has argued that what we get via the S&M characterisation is not the relevant law, per se, but a series of conjunctions of the form $\forall x((Px \,\&\, S_1x) \rightarrow M_1x) \,\&\, \forall x((Px \,\&\, S_2x) \rightarrow M_2x) \ldots$ So, consider the much-used example from physics of bringing a test charge in from infinity to a charged particle. S_1 will be the stimulus of placing the test charge at position r_1, and M_1 will be the relevant manifestation of the charge disposition possessed by the particle, which will either be the force felt or the acceleration experienced, depending on one's analysis of dispositional manifestations. Likewise, S_2 will be the stimulus of placing the test charge at position r_2, and M_2 will be the relevant manifestation of the charge disposition possessed by the particle ... rinse and repeat. However, Vetter insists, a set of conjuncts is not a law – in effect what we have in our example is a set of instances of Coulomb's Law, not the law itself – and the dispositionalist is then faced with an explanatory gap, namely that of accounting for why the conjuncts all have the same form.

Mumford, on the other hand, has argued that if laws 'flow from' or supervene on dispositions, then we can eliminate the former from our metaphysical pantheon (Mumford 2006). Granted, as he notes, there is a tension between the dispositionalist account and the governing feature of laws which non-Humeans tend to emphasise, many philosophers of science would take such eliminativism as effectively a *reductio* of the account itself (although one could moderate this position by insisting that such eliminativism does not imply that we cannot talk of laws in scientific practice, or in philosophical reflection upon that practice, but that such talk should be seen as short-hand for a list of the relevant dispositions – a shift that is curiously akin to the Humean analysis of this talk!).[1] Furthermore, as I shall suggest here, it is, at the very least, not at all clear that dispositionalism can accommodate symmetries in a similar manner and hence the kind of eliminativism that Mumford envisages may founder on this obstacle.

Typically, the laws that are claimed to 'flow from' dispositional properties are understood as causal laws and the causal role of dispositions is explicitly incorporated into the characterisation of this account, as I have indicated above. Immediately one might then doubt whether the account can be extended to the quantum realm, where attributions of causation are famously problematic.[2] More generally, Saatsi

[1]Although it almost goes without saying that the two analyses differ profoundly when it comes to causation!

[2]I'd like to thank Juha Saatsi for pointing this out.

has focussed critically on the claim that a property's 'causal profile', understood as its potential for entering into causal relations with other properties, can be nicely captured on this account in dispositional terms (i.e., via its causal powers; Saatsi 2017). However, setting aside the issue that he also discusses of whether such causal relations have a home in the quantum domain, there is more to charge, to use that example again, than is captured by its causal role (as expressed – setting aside Mumford's and Vetter's concerns – via Coulomb's Law in classical physics). In particular, there is the fact that charge is conserved, something that falls outwith its causal role. It then transpires that such conservation of properties can be related to certain fundamental symmetries (via Noether's Theorem; see Brown and Brading 2003) and in the case of charge, the relevant symmetry has to do with the gauge invariance of the electromagnetic field. Indeed, as Saatsi (2017) points out, it can be argued that much of the dynamical behaviour associated with charge can be attributed to this symmetry in the quantum context. Thus, there are aspects of this property, its behaviour, the way it features in the relevant laws etc., that the standard dispositional account cannot seem to capture.

Relatedly it has been argued that the role of such symmetries in modern physics lies beyond the scope of dispositionalism (French 2014; Cei and French 2014) and what I shall do in the rest of this essay is explore some possible responses to this argument, indicate how these responses run into the sand and then consider how one might draw on certain recent suggestions that go beyond the standard dispositional account in order to appropriately capture the modal features of modern physics.

12.3 Symmetries in Physics

There has been an enormous amount written about symmetries in physics, particularly in the popular science literature. Here's a representative statement from a well-known advocate of their role:

> Nature, like an enemy, seemed intent on concealing from us its master plan. . . . At the same time, we did have a valuable key to nature's secrets. The laws of nature evidently obeyed certain principles of symmetry, whose consequences we could work out and compare with observation, even without a detailed theory of particles and forces. There were symmetries that dictated that certain distinct processes all go at the same rate, and that also dictated the existence of families of distinct particles that all have the same mass. Once we observed such equalities of rates or of masses, we could infer the existence of a symmetry, and this we thought would give us a clearer idea of the further observations that should be made, and of the sort of underlying theories that might or might not be possible. It was like having a spy in the enemy's high command. (Weinberg 2011, p. 1)

Here we have symmetries playing what is essentially a heuristic role in uncovering not only further observations but new theories. This is well-documented in the philosophy of science (see, for example, Post 1971) and numerous examples can be given but let us consider gauge invariance again: this refers to the way in which the Lagrangian of a system – which basically captures the dynamics of that system – remains invariant under a group of transformations, where the 'gauge' denotes certain redundant degrees of freedom of that Lagrangian. Such a group of transformations is represented via the mathematics of group theory and the generator of the group then represents a field. When such a field is quantised, we get kinds of quanta called gauge bosons.

I'll explain what a boson is shortly but if we take electrodynamics, for example, the relevant gauge symmetry group associated with the property of charge is labelled U(1) and the gauge boson that effectively drops out of the requirement to achieve gauge invariance is the familiar photon. This requirement of gauge invariance can then be extended to the other forces in physics and to a significant extent the (in) famous Standard Model of elementary particle physics was constructed on the back of this extension with the weak nuclear force (associated with radioactive decay listed by Mumford above) accommodated via the so-called SU(2) symmetry group associated with isospin (a property of protons and neutrons), and with the strong nuclear force (responsible for binding the nucleus together) associated with the so-called SU(3) symmetry group which operates on the colour property of quarks. Thus the Standard Model is fundamentally a gauge theory, associated with the groups SU(3) x SU(2) x U(1) via which the relevant symmetries can be captured within the theory. With the Higgs boson associated with the breaking of the isospin symmetry of the unified electro-weak force and responsible for the acquisition of mass, we have a complete picture of the fundamental forces, with the exception of gravity of course (how this can be unified with the Standard Model remains an on-going area of intensive research, but certain theories of quantum gravity have been constructed on the basis of gauge invariance, with the graviton as the corresponding gauge boson).

Now, as I said, I shall return to this picture below, but the point I want to emphasise here is the fundamental importance of gauge invariance and, more generally, of symmetry as a heuristic principle in the construction of the Standard Model. Of course, that's all well and good from the perspective of the philosophy of science but what bearing does all this have on the *metaphysics*? If one is an empiricist, constructive or otherwise, one might be tempted to stop there and simply note this heuristic role or perhaps, going a little further, that symmetry is the 'key' to theory (van Fraassen 1989). But if one has realist inclinations, one might be inclined to argue that this role signifies more than this and that the ubiquitous nature of symmetries in modern physics reflects their status as fundamental features of the structure of the world (French 2014). However one conceives of that status, there is the issue of the relationship of such symmetry principles to the familiar laws of physics. Here one can follow Wigner, one of the earliest explorers of the role of symmetries in physics (together with the great mathematician Weyl), who articulated the view that symmetries should be regarded as meta-laws, in a sense, or *constraints* on the more familiar laws of nature (Wigner 2003).

To indicate what is meant here, consider the most fundamental equation of quantum theory, namely Schrödinger's Equation:

$$i\hbar \frac{\partial |\varphi(n_i) >}{\partial t} = H|\varphi(n_i) >$$

where H denotes a specific Hamiltonian (taken out of another toolbox!) and the n_i denote the state-independent properties that identify the kind of particle involved. It is worth noting that to describe this as a law is not to do it justice. In effect Schrödinger's equation is a kind of 'uber-law' that operates at a very high level of generality. We only get laws applicable to specific systems once we plug in the relevant Hamiltonian (Cartwright 1999 refers to this as 'fitting out' the equation to yield a concrete model of a physical system). Thus, Schrödinger's equation acts as a kind of structural constraint itself on the laws of specific systems.

To attribute a property to a particle obeying one of these specific laws, the operator for that property must commute with the corresponding Hamiltonian (in the sense, to put it a little crudely, that it doesn't matter whether one applies that operator first to the state function describing the system or the Hamiltonian). Sets of such operators may form a group which represents a symmetry of the system (see McKenzie forthcoming, for a clear presentation). By virtue of the requirement that these operators must commute with the relevant Hamiltonian, the symmetries can be viewed as acting as constraints on particular Hamiltonians and their associated particular laws. Hence the specific laws obtained from Schrödinger's equation are constrained in ways that do not sit comfortably with dispositionalist accounts (see Cei and French 2014). To flesh this out, let's consider a specific example: Permutation Symmetry, which plays an absolutely fundamental role in quantum physics.

Consider the possible arrangements that arise from distributing two particles between two boxes, representing states. In classical 'Maxwell-Boltzmann' statistics we obtain four possible arrangements: both particles in the left-hand box, both in the right hand and one in each, the latter arrangement being counted twice because the particles can be permuted, and such permutations are typically taken to have physical significance in this context. This counting is fundamentally important: according to Maxwell-Boltzmann statistics the probability of obtaining both particles in the left-hand box, say, is ¼ and it is this assignment of probabilities that forms the heart of classical statistical mechanics and grounds the latter's underpinning of thermodynamics. In quantum statistics however, the 'one particle in each box' arrangement is only counted once; that is, to put it a little crudely, permutations are not counted in the statistics, so quantum theory is said to be 'permutation invariant' – we then obtain two forms: Bose-Einstein statistics, which allows both particles to be in the same box and hence for which the probability of both in the left hand box is now 1/3; and Fermi-Dirac statistics, which excludes both particles being in the same box and hence only allows the one particle in each box arrangement (see, for example, French and Krause 2006).

Let me say two further things about this piece of physics. First, the distinction between Bose-Einstein and Fermi-Dirac statistics manifested in the difference between the above probabilities is absolutely fundamental. Particles obeying the former – bosons – exhibit a statistical tendency to cluster together, something that is exhibited most strikingly in the phenomenon of 'Bose-Einstein condensation' (first produced in the laboratory in 1995 and for which achievement the Nobel prize was awarded in 2001; see http://en.wikipedia.org/wiki/Bose–Einstein_condensate) but the ability to pack bosons into the same state also gives us lasers – at the heart of so much modern technology of course! Fermions, on the other hand, cannot occupy the same overall state (where this is taken to encompass all the relevant properties – so two fermions can occupy the same energy state but only as long as they have different spins, for example). Thus, this permutation invariance is what lies behind the famous Pauli Exclusion Principle, which in turn accounts for the Periodic Table and thus, contentiously perhaps, for chemistry. More generally, it is this behaviour that results in many of the well-known and 'everyday' properties of matter, such as solidity for example. Thus, fermions, such as electrons, are typically regarded as 'material', whereas bosons are taken to be the 'force-carriers' between such material particles.

Secondly, in mathematical terms, these two forms of quantum statistics correspond to different irreducible representations of the Permutation Group, which mathematically encodes this symmetry. (A group is a set of elements together with an operation that combines those elements, according to certain axioms, and a representation in effect renders the group more concrete by representing it in terms of (linear) transformations in some vector space, such as the Hilbert space of quantum mechanics.) The application of the mathematics of group theory to quantum mechanics was of major theoretical significance, both in terms of providing a further set of foundations for the theory and also simply as a way of yielding solutions to otherwise difficult to crack dynamical problems (see, for example, French 2014, ch. 4 and references therein). The bottom line as far as we are concerned here, is that in order to be physically acceptable, any Hamiltonian must commute with the permutation operator which combines the elements in the Permutation Group, where these elements represent quantum entities. This is a fundamental demand which crucially constrains our quantum theoretical description of an assembly of such entities and yields these mathematical representations corresponding to the two most fundamental kinds of particles, with radically different statistical behaviour.

However, it turns out that these are not the only kinds of statistics that are possible. If we consider more than two particles, other forms are compatible with the above constraint, forms that also correspond to different representations of the Permutation Group (for a summary of the relevant history, see French and Krause 2006; or for more general details, French 2014). These forms are generically known as 'parastatistics' and although they were known to the likes of Dirac back in the 1930s, it wasn't until the early 1960s that they came to be more fully explored as theoretical possibilities. Indeed, it was hypothesised at one point that quarks should be regarded as paraparticles. But this idea subsequently fell out of favour as the relevant statistical behaviour of the particles concerned was accounted for in terms of a new quantum number – 'colour' – together with standard quantum statistics, as outlined above, leading to the development of the theory of quantum chromodynamics as a theory of the strong nuclear force.

Now, there is much more to say, of course, but my aim here is just to illustrate how symmetries, as exemplified via the Permutation Group, impose fundamental constraints on theories, such as quantum mechanics. The question now is, how can dispositionalism accommodate these kinds of constraints?

12.4 The Dispositionalist Responds

One option is to take such symmetry principles to be nothing more than "pseudo-laws", or false constraints, to be written out of our scientific world-view (Bird 2007, p. 214). As well as removing this obstacle to the extension of dispositionalism in general, it would also allow Mumford, for example, to pursue his eliminativism, as mentioned above. However, given the importance of such symmetry principles in modern physics and their role within the Standard Model in particular, I shall dismiss this response as tantamount to a counsel of despair (Livanios 2010; see also Cei and French 2014).

Another option might be to shift the 'seat' of the disposition, away from objects such as particles to, in this case, the world as a whole (see Bigelow et al. 1992; also Bird 2007, p. 213). Now, I'll come back to this issue of shifting the 'seat' (see Sect. 12.5), but let me just remark here that it raises the further issue whether it is of the essence of dispositionalism that it be 'object oriented'. In other words, must it be the case that only objects (however understood) can possess powers or dispositions? Insofar as dispositions arise from a second order analysis of properties (at least according to the Dispositional Identity Thesis) one might be inclined to insist that anything, object or no, that could be said to possess properties could then be said to have or possess dispositions. And if symmetries are seen as (relational) properties of laws, then one could argue that the latter could be regarded as the seat of the dispositions that give rise to symmetries. Whether such a view is still entitled to the name of dispositionalism is a further matter for discussion.

Here the issue is whether we can shift the seat to 'the world', still considered as an object. Now that latter point might be contested (see van Fraassen 1995), although monists (at least of a certain stripe; see for example Horgan and Potrc 2008) of course will insist that it is the only object! On such a view, the 'usual', local dispositions associated with properties, such as charge etc., would presumably be considered to be derivative of 'global' dispositions associated with the world-as-object. If one were unhappy with treating the world itself as an object, one might still retain the global character of the view within the kind of structuralist account that I favour (French 2014). I shall return to this in Sect. 12.7, below.

But if we set such moves aside for now, the obvious worry is that grounding the dispositions to be associated with symmetries in 'the world', considered as an object in itself, seems ad hoc (Bird 2007, Livanios 2010.). Furthermore, and perhaps even more damagingly, it is not clear – on the Dispositional Identity Thesis – what the relevant property would be that would be decomposed into the appropriate dispositions. If that property is taken to be 'being the world' then we would have one property associated with myriad different dispositions. At the very least, the value of the dispositional analysis would appear to be cheapened by such a move. Alternatively, if the relevant properties of the world are taken to be those associated with the relevant laws, then one would have to wonder what the invocation of the world-as-object is doing when it is the laws that are the actual seats of the relevant dispositions. Finally – and again, this is an issue we shall return to – if it is the world (that is, the universe!) that is taken to possess, in whatever sense, the relevant dispositions, with the symmetries as manifestations, then one has to wonder what the associated stimuli could be. Clearly, in this case, shifting the seat of the dispositions but retaining an object-oriented approach undermines the S&M analysis.

Alternatively, one might try tweaking that S&M characterisation. Thus Chakravartty and Heil have suggested that the manifestations of dispositions may be relational in nature. Chakravartty (2007) has argued that the first-order causal properties of scientific objects by which we detect such objects, should be understood in terms of dispositions for specific relations which comprise the concrete structures about which we should be realists. Similarly, Heil (2005) argues that the manifestation of a disposition is a mutual manifestation of reciprocal disposition partners, so that rather than thinking of the S&M characterisation in terms of a chain, we should conceive it as yielding a kind of relational 'net', which might then be useful compared with Chakravartty's concrete structures. Heil offers the classic example of the dissolution of salt in water: we can see this in terms of the water being the stimulus for the dissolution of the salt or the salt being the stimulus for the manifestation of the saltiness of the water. As Chakravartty makes clear, this relational conception appropriately captures the relevant features of scientific properties and associated laws, although when it comes to the mutual manifestation of reciprocal disposition partners Reutlinger has dismissed this as 'just a fancy label for an uncontroversial fact' (Reutlinger 2013).

These are interesting moves, and certainly go some way towards accommodating various features of scientific theories (see for example Chakravartty 2013 and the discussion there). However, with symmetries regarded as constraints, or meta-laws, or more generally as relations between relations, it is difficult to see how they can be accommodated even by generalising the notion of a manifestation in the ways that Chakravartty and Heil suggest. The issue is how we get up to the 'next level' as it were, from manifestations as relational to symmetries as relations between relations.

One option, again, would be to shift the seat of the dispositions from the objects to the laws. However, it remains difficult to see how the S&M analysis would apply – again, what would be the stimuli in such cases leading the *laws* to manifest the relevant symmetries? There is a sense in which the reason the standard S&M analysis 'fits' an object oriented ontological stance is that with objects taken to be (minimally perhaps) spatio-temporally delineated in some manner, one can pretty straightforwardly conceive of external stimuli, such as other objects, or forces etc. In the case of laws, as standardly understood, it is difficult to conceive of such stimuli – if the laws are taken to be universal (which is why I specified 'standardly understood') what would be external to them? Objects don't affect laws and neither do forces – indeed, laws just don't seem to be the kinds of 'things' that could be affected that way (again, at least not as standardly understood). Of course, one can find non-standard views that take laws to be non-universal in nature or regarded as nomological entities that one would not normally take to be such (such as the wave-function in quantum mechanics, for example, something else we shall touch on again below). But again, it is difficult to see what would count as the relevant stimuli in such cases, leading to the manifestation of symmetries. One might appeal to some sort of 'multiverse' account and somehow argue that whatever it was that generated this particular universe with its associated laws and symmetries could count as the stimulus of the former generating the latter, but that would be to ride off into distant realms of speculation.

Another option would be to package up the laws and symmetries together, in the sense that the manifestation, under the S&M analysis, should be understood not just as relational, in the way that Chakravartty and Heil suggest, but as second-order relational in that it includes the relations between relations that come to be categorised as symmetries. In other words, the disposition yields the relevant law and symmetry as a bundle. An attractive feature of such a proposal is that it might explain why symmetries and laws are bound so tightly together – they are so because they both derive from the same source, namely the relevant disposition.

However, at the very least this suggestion needs fleshing out to show how it would actually pan out within the dispositionalist framework. Again, consider the S&M analysis: at the heart of Vetter's criticism of it (see above Sect. 12.2, pp. 181f.) is the claim that the best that we get is a bunch of instances of the form 'if we bring a charge up to the test charge at such-and-such a distance, it experiences such-and-such a force (or acceleration)' and so on. What we don't get is Coulomb's Law itself, or at least not as standardly understood as a universal generalisation. We are then left with an explanatory gap, something which alternative accounts of laws can exploit (see, again, French 2014, ch. 9 for discussion).

If we can't even get to laws from instances of S and M, then it remains completely unclear how we get up to the next level, to symmetries. Even if one were to insist that the appropriate manifestation is not just the experienced force, or acceleration, but the force or acceleration as conditioned by what we come to identify as the associated symmetry (in the case of charge, represented by the U(1) group), this still does not take us 'up' to the universal nature of such symmetries, for the same reasons as Vetter identifies.

There is also the issue of whether this analysis, even if successful, could capture the way in which symmetries act as constraints. Insofar as this is discussed, it is typically taken to be a 'top-down' kind of constraint; indeed, Wigner understood symmetries to be 'meta-laws' in a sense. Now of course to insist on that 'top-down' nature would be to beg the question against dispositionalism, in just the same way as insisting that laws govern properties would. As Mumford (2006) has astutely noted, the latter conflicts with the reductive, bottom-up approach of dispositionalism and setting aside the issue of whether his eliminativism can be applied to symmetries as well, one can at least acknowledge that this constraining feature of symmetries must be reconceived in this context. In essence, the dispositionalist must maintain that it is of the nature of the dispositions in terms of which charge, for example, can be analysed that the laws that 'flow' from or supervene upon them are then constrained by these further features. Setting aside Vetter's concerns as to whether this supervenience even holds in the first place, the constraint would have to be conceived of as bound up in the manifestation, so that what appears to be top-down is actually exercising itself from the bottom up, as it were.

In the case of symmetries associated with particular properties, such as U(1) and charge, or SU(2) and isospin, that seems a possible way forward at least, although, as I have said, it has yet to be developed in any significant form (also see Bauer 2011) and it may be metaphysically more straightforward to simply drop the S&M analysis entirely. I shall return to this idea in Sect. 12.7. But what about permutation symmetry? This seems to impose a top-down constraint that applies 'across the board', as it were, and cannot be folded into any dispositionalist reduction of a particular property. Or can it?

12.5 From Statistics to Spin

It turns out that there is a connection between the statistical behaviour of elementary particles and the quantum property known as 'spin' (enfolded within the extension of dispositionalism by Mumford, for example, as we have seen): particles that obey Bose-Einstein statistics have integral spin and particles that obey Fermi-Dirac statistics have half-integral spin. This connection is underpinned by the 'Spin-Statistics Theorem', a proof of which was famously given by Pauli, using considerations which draw from Einstein's Special Theory of Relativity. So, if we see spin as that property whose identity is given in terms of the disposition for particles to behave in a certain statistical way, we might be able to bring permutation symmetry within the dispositionalist purview.

Now, of course, spin is an interesting property in that it has no classical analogue. Although it is associated with the 'intrinsic' angular momentum of the particles and has the same units as classical angular momentum (the Joule-second), one should not think of the particle as literally spinning like a little top (the 'direction' of the spin is not like that of a classical vector quantity for example). Nevertheless, the recent exploitation of spin in electronic devices – leading to the field of 'spintronics' (see https://en.wikipedia.org/wiki/Spintronics) – establishes it as a measurable and technologically usable property – just like charge – that should in principle be capable of being understood from a dispositionalist perspective.

However, even granted all of that, it turns out that symmetry re-enters the picture in a way that undermines the attempt to bring spin within dispositionalism's grasp. One of the virtues of Pauli's proof of the theorem is that by calling on relativity theory it meshes with what is generally understood to be spin's relativistic nature. And that nature can be summed up in the pithy statement that the property effectively 'drops out of' another group – the Poincaré group, which is the symmetry group of relativistic field theory. More specifically, it is the group that applies to Minkowski space-time – the space-time of Special Relativity – and it turns out that the irreducible representations of *this* group yield a classification of all elementary particles, with these representations indexed or characterised by mass and spin (the invariants of the group). Thus, even if we were to 'ground' the particle statistics generated by Permutation symmetry in the property of spin, and analyse the latter dispositionally à la charge, that analysis would still have to deal with symmetry, now in the guise of the Poincaré group.

Now, there are two things to note here. First of all, there is the issue of how we should regard spin qua property. If we adopt the apparently plausible view that our metaphysical analysis of the nature of a given property should at the very least touch base with the theoretical and experimental practices by which we gain epistemic access to that property, then we need to confront the fact that although the rise of spintronics demonstrates the causal impact of spin, its existence is effectively guaranteed by the relevant symmetry as expressed via the Poincaré group. Thus as Morrison emphasises, in her extensive study of the history and physics of spin, "[o]ur current understanding of spin seems to depend primarily on its group theoretical description" (2007, p. 552). How that group theoretic feature should be captured ontologically remains a contentious issue – Morrison herself argues that it implies that spin should be regarded as a "hybrid" mathematical and physical property (2007, for criticism see French 2015). I shall not pursue this issue here but simply note her point that: "[p]art of the difficulty with attempts to generate a physical notion of spin concerns the way the electron is pictured in the hydrogen atom as a quantum mechanical object" (Morrison 2007, p. 554). Thus the involvement of symmetry via the group theoretic characterisation of spin suggests that we should drop the picture of it as a property that is associated with objects in the usual way.

Secondly, one can again make the case that the relevant symmetry should be seen as a kind of 'meta-law' or top-down constraint. Thus Lange (2013) has argued that such a characterisation explains why the Lorentz transformations would still have been true, even if the relevant force laws had been different. The former have a greater modal strength than the latter in the sense that we can conceive of relativistic worlds in which the symmetries hold (otherwise they wouldn't be relativistic!) but the relevant laws don't. Thus the dispositionalist, seeking to escape the constraining effect of Permutation Symmetry by appealing to the Spin-Statistics Theorem, finds herself hopping out of that particular frying pan and into the fire!

Nevertheless, granted that spin should be seen as a consequence (in some sense) of space-time symmetry, there might still be options available to the dispositionalist. Let us explore them again *via* the question of what the 'seat' of the relevant disposition might be.

12.6 Seating Spin

Thus the first option would be to take that seat to be the particle, and to regard spin as associated with the disposition that manifests Poincaré symmetry, where the manifestation here is obviously relational, in the manner articulated by Chakravartty (2007) and Heil (2005) perhaps. Now this option would appear to be blocked by what we have just said regarding the way spin 'drops out' of the symmetry. However, one could try to turn the explanatory arrow the other way around and insist that spin can be associated with a certain kind of dynamical behaviour of particles that we then codify in terms of the Lorentz transformations and the Poincaré group. This would be to reject the latter as a constraint and to argue that insofar as it is taken to represent the structure of relativistic space-time, that structure, as represented geometrically, is essentially a manifestation of the dynamical behaviour of the particles and thus of the force laws.

This would appear to be quite a radical line to take but it has been taken, most notably by Brown (2005). He argues that the structure of relativistic space-time should indeed be seen as "[. . .] a codification of certain key aspects of the behaviour of particles" (ibid., pp. 24–25) and the dispositionalist could hitch her wagon to Brown's account and argue that spin, as a property, features in certain dynamical laws which manifest or express Lorentz invariance as a result of which relativistic space-time has the structure that it does. However, the nature of the explanation then becomes unclear and in particular how it is that the symmetry, qua a feature of the relevant laws, results in the relevant geometric structure of space-time (see Skow 2006). And, of course, the dispositionalist still has to account for that feature within the framework of the S & M characterisation. Finally, as a possible 'seat' of these obscure dispositions, particles themselves are notoriously problematic in this physical context, despite physicists' use of the term, with a number of well-known arguments to the effect that the notion of particle, as usually conceived, cannot be sustained within relativistic quantum field theory (see, for example, Bain 2011).

An alternative option would be to shift to a field-theoretic ontology and insist that it is the quantum fields that offer an appropriate seat for the dispositions, with spin understood as a field property. Now insofar as fields are global entities, we might see this as akin to the previous option of taking 'the world' as this seat, with the crucial difference that they may be fewer reservations about the appropriateness of fields in this regard as compared to 'the world'.

Of course, there is then the further issue of the relation between the field *qua* global entity and relativistic space-time, structured via the Poincaré group. Two further options arise, depending on how one understands fields (see French and Krause 2006, pp. 51–64). One is to conceive of them as global substances, spread out, jelly-like, across space-time. In this case, if the field is to be the seat of the disposition in terms of which spin is understood such that it can be associated with the Poincaré group, something is going to have to be said about the relationship between that global entity and space-time; however, it is not clear what. The alternative – and perhaps more widely held – option is to understand fields in terms of assignments of field quantities or properties to space-time points (where these assignments will be appropriately 'smeared' in the case of quantum fields). In this case, the field itself is not substantival but simply represents the way in which properties are distributed across the whole of space-time.

Following this latter option obviously shifts the seat of the relevant dispositions from the field *per se* to space-time. Again, there are further options. One is to take the relevant seat to be space-time *qua* entity as a whole, bringing us back towards a monistic position with similar problems in ascribing property-based dispositions on a global scale. Recall: we are not ascribing Poincaré symmetry *qua* disposition to space-time as a global entity but rather are so assigning spin and other properties as expressed in the usual force laws from which the symmetries will emerge in some fashion. Without the form of ontological pointillism that particular objects provide (just to alliterate a little) it is difficult to make sense of the manner in which these various dispositions can be supported. And there is still the problem of articulating any sense of external stimulus in this context.

Fortunately, there is the further alternative of taking space-time to be ontologically composed of a manifold of points, which supports various structures, including that expressed via the Poincaré group. The field assignments are then to these points or regions composed of them. And the seat of the relevant dispositions would be these particular space-time points themselves. One might then be able to elaborate a physics-based form of dispositionalism according to which both space-time structure and properties such as spin can be regarded – via the mediation of field theory – in terms of dispositions possessed by the points of relativistic space-time.

Now, of course, this would be to commit dispositionalism to a form of substantivalism with regard to space-time, which many have found problematic, as is well known (for a useful survey see Norton 2019). Certainly, as objects, in a sense, the points of Minkowski space-time are not really on a par with the kinds of objects that dispositionalists usually latch onto. Although these points, or the manifold that they constitute, support the kinds of structures indicated above and thus in a sense (disputed by the likes of Brown, of course) could be said to causally constrain physical processes (so the 'light-cone structure' prevents processes from accelerating past the speed of light), there is a lack of causal reciprocity in that the space-time structure itself is not affected by these processes.

This reciprocity is restored in General Relativity, but unfortunately, in that context, there are well-known arguments that have been taken to rule out the above straightforward ontological understanding of space-time as constituted by a manifold of points supporting various structures (namely the so-called 'hole' argument which shows that on such an understanding the points can be shifted about, creating a 'hole' which is compatible with all the relevant dynamics, leading to a kind of inadmissible ontological indeterminacy; again for an accessible introduction, see Norton 2019). Those of a substantivalist inclination have responded to such arguments, in particular noting the assumption that the points have a kind of primitive identity and that the arguments dissolve if this is dropped (Pooley 2006).

According to the 'sophisticated substantivalism' that results, the space-time points become no more than placeholders to support the relevant structure that embodies the physics (so again we move towards a kind of structuralism).

At this point, the question becomes: how can that which is only a placeholder for relevant space-time structure(s) be regarded as the seat of the dispositions from which flow (somehow) or on which supervene (again, somehow) the symmetries constituting that structure? At best this particular seat is very thin indeed.

But things get even worse for the dispositionalist. One might argue that these space-time symmetries are, at least, physically implementable: one can translate objects through space, or rotate them or give them a swift kick and thus a boost (and if the kick is hard enough, the boost may be significant enough to warrant deployment of the Lorentz transformations!). Thus one could form the following subjunctive conditionals:

If *P* is a real (determinate) physical property of *x* and were *x* to be translated through space, then *x* would remain *P*

If *P* is a real (determinate) physical property of *x* and were *x* to be rotated through space, then *x* would remain *P*

⋮

If *P* is a real (determinate) physical property of *x* and were *x* to be given a Lorentz 'boost', then *x* would remain *P*

Taking them in conjunction, one could then form the more general subjunctive conditional: If *P* is a real (determinate) physical property of *x*, and were *x* to be Poincaré transformed, then *x* would remain *P* (see McKenzie forthcoming).

Thus one can envisage this kind of symmetry being shoe-horned into the S&M analysis by taking the translations, rotations and boosts to be the stimuli and the invariance of the given property as the manifestation. This would then mesh with Brown's approach to relativity theory, indicated above, whereby the focus is on the objects, or more generally, systems and their dynamical behaviour, with the relevant symmetries arising out of that.

However, concerns arise with regard to the subjunctive element in the above formulations: what does it mean to say '*were* x to be Poincaré transformed'? It seems straightforward to understand how an object might or might not be translated through space or boosted or whatever, but there's a jump from this to a similar understanding about subjecting the given object to the Poincaré transformations, *where these are understood to be representations of certain features of the structure of space-time*. With the (local) space-time understood to be Minkowskian, talk of *x* being 'transformed' seems to attribute power inappropriately – as if *x* might not be. But given that *x* is situated within this particular space-time structure it is not as if the Poincaré symmetry can be turned off somehow; in other words, from this perspective, it is occurrent. Of course, this is the perspective of the standard, non-Brownian view, but even if one were to adopt Brown's account, there are symmetries that are not physically implementable in this manner.

12.7 Internal Symmetries

These are the so-called 'internal symmetries' (some of which have already been mentioned with regard to the Standard Model). So, if one kind of particle, defined by a certain set of quantum numbers, evolves or interacts in accordance with some Hamiltonian, then there is another kind of particle, defined by a different set of quantum numbers, that also does so. Thus families of particles can be formed, or 'multiplets', associated with different groups and representing different forces, as indicated earlier:

Electromagnetic: U(1)
Weak nuclear: SU(2)
Strong nuclear: SU(3)

And as was noted above, the combination SU(3) × SU(2) × U(1) forms the mathematical basis of the Standard Model.

So, in the case of the weak nuclear force, responsible for radioactive decay (as we have seen, explicitly embraced by the dispositionalist), the associated quantum number is known as 'weak isospin', which is conserved in interactions involving this force, with the relevant group SU(2). Isospin was originally introduced in the context of the strong nuclear force as a way of accounting for certain symmetries associated with the then newly discovered neutron. In particular, it was noted that the mass of the neutron was comparatively close to that of the proton and the strength of the strong interaction did not depend on whether the particles were protons or neutrons (the fact that the masses are not the same means that the symmetry is broken). Thus, ignoring the mass difference, one could treat the proton and neutron as two different states of the same entity, the nucleon, each with different isospin. Isospin was originally modelled on the mathematical representation of spin and was subsequently extended to quarks (the up and down quarks also have very similar masses), thus requiring an enlargement of the associated group to SU(3).

Now, what is important is that the transformations represented by these groups do not 'take place' in physical space, as in the case of the Poincaré group; rather they are represented via highly abstract mathematical spaces that have no physical counterparts. In particular, these 'internal' symmetries are not physically implementable and thus no subjunctive conditional can be formed as in the Poincaré case. In effect, the symmetries are always 'on' and hence, again, are occurrent (see, again, McKenzie forthcoming). Now, even if we were to accept the above shoe horning of the Poincaré group, *these further symmetries that lie at the heart of the Standard Model simply cannot be accommodated by the S&M picture.*

Enough already. Let's recall that we started off down this branching path in an attempt to avoid the constraint imposed by Permutation Symmetry by crossing the bridge offered by the Spin-Statistics Theorem and grounding the statistical behaviour of particles in their spin. But spin in its turn appears to be grounded in a further symmetry, as represented via the Poincaré group. Attempts to ground that symmetry in some further particular, either global or local, appear to be fruitless and the Stimulus and Manifestation account looks simply inappropriate when it comes to these space-time symmetries but even more so when we take on board the 'internal' symmetries of the Standard Model. So perhaps we should just acknowledge the constraining nature of these symmetries and attempt to adapt dispositionalism accordingly. The problem, as we have repeatedly noted, is not just with regard to what would count as the seat of the disposition – we seem to have reached the point where the best we can say is that the space-time structure itself is that seat, in some sense – but also with the Stimulus and Manifestation framework within which dispositionalism is standardly presented. Again, even if we can articulate a sense in which these constraining symmetries can function, metaphysically, as seats of dispositions, it is unclear, or perhaps even impossible, to see how they can be appropriately stimulated!

Perhaps, within the framework of the 'toolbox' approach, as introduced at the beginning of this essay, there are devices or moves in recent developments and extensions of dispositionalism that we can appropriate to help capture the above kinds of symmetries and thereby enfold them within a modal metaphysics.

12.8 Taking Up Some Tools

As an example of such a move, consider Mumford and Anjum's reconception of the notion of a stimulus as the pragmatic designation of a contributory power, where such powers should be regarded in terms of *sui generis* tendencies (Mumford and Anjum 2011). Powers, on this view, should be seen to be not so much stimulated as 'unleashed'. Consider the example of a pool of petrol in an oxygen-rich environment: the spark that ignites the conflagration should not be seen as a stimulus per se but as a further contributory power. A given effect or manifestation may then be the result of many such powers operating at once, and Mumford and Anjum give an analogy with the vector addition rule for forces (for criticism see Glynn 2012; Pechlivanidi and Psillos, Chap. 8, this volume). Obviously, such an analogy is going to be inappropriate in the case of symmetries, since these are not 'force-like' and not subject to the same kind of rule, but more importantly, it is difficult to see what the contributory power would be in the situations we have considered here or how a symmetry could be regarded as 'unleashed'.

Cartwright also eschews the standard dispositional analysis and in discussion has suggested that the 'seat' of the relevant powers could be seen as the quantum state itself, which encodes the properties, laws and symmetries in one package. One can see a similar idea in Esfeld et al.'s recent work where they argue that the particle configuration as a whole instantiates the dynamical, dispositional property that determines the temporal development of the system and which is represented by the universal wave function (Esfeld et al. 2017). Again, there is no question of there being any stimulus or external power involved – things simply unfold according to one disposition that enfolds everything. Now there is a lot to say about this sort of suggestion, but the immediate worry is that it seems an ad hoc response and, furthermore, loses whatever advantages the dispositional account enjoyed in terms of metaphysical explanation via the S&M analysis.

Here's a further suggestion, in Cartwrightian vein: Hüttemann argues that all laws should be regarded as default laws, in the sense that they 'tell us what happens if nothing interferes'. Causes should then be understood as interfering factors or antidotes that explain why the dispositions ascribed by the laws fail to be (completely) manifested. Such a view allows for partial manifestations, as in the example of a crystal, for which the presence of an impurity would be the interfering factor. Thus, even if there is no complete manifestation, we can still have evidence for a particular disposition (Hüttemann 2009). On this view, the relevant dispositions are continually manifesting and one might regard symmetries likewise. Indeed, one could then construe the kinds of 'internal' symmetries involved in the Standard Model as only partially manifested due to symmetry breaking (as in the case of the mass differences between the 'up' and 'down' quarks which breaks the SU(3) symmetry). However, there are no causal interfering factors in this case, so it is not clear that as it stands this sort of approach could be said to apply to symmetries. Of course, one might envisage an extension of this approach that drops the causal aspect but even so, it remains unclear what might count as the relevant interfering factor.

One might have similar worries about Vetter's suggestion that we should drop the S&M analysis entirely, particularly the S side, and take dispositions to be individuated by the manifestation conditions alone (Vetter 2014, 2015). Of course, as she recognises, this effectively divorces her account from the usual counterfactual analysis (the clue is in the title of her paper!). It also thereby again loses the apparent advantages of standard dispositionalism in explaining or grounding causation and the necessity of laws. However, Vetter argues that these should be seen as a separate 'ingredient' that may or may not be part of the manifestation, depending on the circumstances and the nature of the disposition. More positively she draws on the ways we talk about possibility in our everyday language to argue that dispositional ascriptions are naturally paraphrased via expressions of possibility, the localised counterpart of which is potentiality (2015, p. 23). Thus the relevant sense of modality is best characterised (to a first approximation) by '*x* can *M*' (where this should be understood as graded and context-sensitive).

As Vetter herself notes, by losing the counterfactual analysis and dropping causation, this account appears to be precluded from being applied to scientific properties. But of course, as we all well know, causation is problematic, to say the least, at the level of fundamental physics. Indeed, Nolan, for example, has sought to accommodate this by arguing that certain dispositions are non-causal and drawing on features of quantum physics such as the infamous EPR correlations, particle creation, radioactive decay and so on (Nolan 2015). No doubt, as he would acknowledge, more needs to be said here. So, although radioactive decay and particle creation appear to be non-causal and random, they fall within the framework of the Standard Model, as already noted, and are subject to the kinds of considerations we have indicated above. As for the EPR correlations, although it is certainly true that they face well-known problems being accommodated within a causal framework (Bell's Theorem comes into play here), it is certainly not at all clear how they might be regarded in terms of, or a as a result of, some kind of disposition. Consider: Teller famously argued that one could give a metaphysical explication of these correlations in terms of non-supervenient relations (Teller 1986). If the seat of the relevant disposition is taken to be the quantum objects – perhaps the most obvious choice – then the usual account of the relational manifestations 'flowing from' or 'supervening on' these dispositions is going to be blocked. Furthermore, the very concept of 'object' faces considerable metaphysical problems in the quantum domain (French and Krause 2006) and as in the space-time case, the best we can hope for is a 'thin' notion of object as placeholder for the relevant structures, or in this case, the relevant correlations – and again that seems a very weak peg on which to hang one's dispositional hat!

Returning to Vetter's account, she herself acknowledges that it cannot be applied to fundamental properties in physics. But I think she is being too quick here. Certainly, it cannot be applied to classical properties, where the standard dispositional analysis gets some traction. But we've already seen that this analysis needs to be given up and, in particular, the characterisation in terms of stimuli abandoned. Suppose we were to follow her suggestion, drop the S, and allow for laws and symmetries to be aspects of the manifestation of a graded kind of possibility? Vetter still retains objects as the seat of her localised possibilities, or potentialities, but let us drop those as well, since they are too thin to bear the metaphysical weight. Instead, I would urge that we should take as our seat that which is doing all the physical work, namely the relevant physical structure – that of Minkowski space-time in the case of Special Relativity and that as presented via group theory in the Standard Model (French 2014; McKenzie 2013). Then we have something like this: our seat, not of dispositions (since without the conditional analysis we can hardly call this a dispositional account) but of possibility, is the physical structure, with no relevant stimuli, no S&M analysis and thus no subjunctive conditionals. However, there is individuation via the relevant manifestation (including, in a reversal of the usual metaphysical dependence, properties like spin, mass . . .) and an understanding of Vetter's '*x* can *M*' can be given as 'the structure contains certain possibilities', or possesses certain potentialities. The example would be that of Permutation Symmetry, which contains more possibilities than the Bose-Einstein and Fermi-Dirac statistics that we observe in this the actual world, and thus there is the potential – fleetingly realised in the 1960s – for non-standard statistics to be manifested (Vetter herself prefers what she calls a "piecemeal" approach and rejects this kind of "catch-all" view (2015, pp. 259–263), although I would offer the above as the "new conception of the world as a structured object" (ibid., p. 261) that she feels the catch-all view must provide if it is to be plausible).

Elsewhere I have claimed that we have to accept a kind of primitive modality associated with physical structures of this form (French 2014) but as with all invocations of primitiveness, such a conception seems to suffer in comparison with the richer metaphysical details of an account such as dispositionalism. However, dispositionalism, at least as standardly characterised in terms of the S&M analysis, seems to be ruled out by modern physics and some may see this as a further indictment of current metaphysical reasoning. Yet that would be too quick, as the latter may still provide the tools by means of which we can accommodate symmetries and the like. I've indicated one such tool, namely Vetter's notion of 'potentiality', but there are others which may also be deployed, offering some hope for a naturalised metaphysics that does not require us to build new frameworks from scratch. Perhaps by drawing on these metaphysical tools, we can indeed redress the balance and arrive at a view that can accommodate quarks as well as vases.

Acknowledgements Earlier and different versions of this chapter were presented at the work-in-progress seminar of the Centre for Mind and Metaphysics of the School of Philosophy, Religion and History of Science, University of Leeds, at the Philosophy Seminar of Durham University, the XI International Ontology Congress, San Sebastián, Spain and also at the conference, 'Real Possibilities, Real Absences', organised by Kristina Engelhard and David Hommen at the University of Köln. I am grateful to all the participants but especially Nancy Cartwright, Anjan Chakravartty, Daniel Elstein, Andreas Hüttemann, Kerry McKenzie, Margie Morrison, Daniel Nolan, Juha Saatsi, Pete Vickers, Robbie Williams and Richard Woodward for all their helpful comments and suggestions. Finally, I would particularly like to thank Anne Sophie Meincke for her specific comments on my contribution and of course for the initial invitation.

References

Bain, Jonathan. 2011. CPT Invariance, the Spin-Statistics Theorem and the Ontology of Relativistic Quantum Field Theory. *Erkenntnis* 78: 797–821.

Bauer, William A. 2011. An argument for the Extrinsic Grounding of Mass. *Erkenntnis* 74: 81–99.

Bigelow, John, Brian Ellis, and Caroline Lierse. 1992. The World as One of Kind: Natural Necessity and Laws of Nature. *The British Journal for the Philosophy of Science* 43: 371–388.

Bird, Alexander. 2007. *Nature's Metaphysics: Laws and Properties*. Oxford: Oxford University Press.

Brading, Katherine, and Harvey R. Brown. 2003. Symmetries and Noether's Theorems. In *Symmetries in Physics: Philosophical Reflections*, ed. K. Brading and E. Castellani, 89–109. Cambridge: Cambridge University Press.

Brown, Harvey R. 2005. *Physical Relativity: Space-time Structure from a Dynamical Perspective*. Oxford: Oxford University Press.

Callender, Craig. 2011. Philosophy of Science and Metaphysics. In *The Continuum Companion to the Philosophy of Science*, ed. S. French and J. Saatsi, 33–54. London: Continuum.

Cartwright, Nancy. 1999. *The Dappled World: A Study of the Boundaries of Science*. Cambridge: Cambridge University Press.

Cei, Angelo, and Steven French. 2014. Getting Away from Governance: A Structuralist Approach to Laws and Symmetries. *Méthode – Analytic Perspectives* 4: 25–48. http://citeseerx.ist.psu.edu/viewdoc/download?doi=10.1.1.840.665&rep=rep1&type=pdf. Last accessed on 9 Nov 2019.

Chakravartty, Anjan. 2007. *The Metaphysics of Science*. Cambridge: Cambridge University Press.

———. 2013. Realism in the Desert and in the Jungle: Reply to French, Ghins, and Psillos. *Erkenntnis* 78: 39–58.

Esfeld, Michael, Dustin Lazarovici, Vincent Lam, and Mario Hubert. 2017. The Physics and Metaphysics of Primitive Stuff. *The British Journal for the Philosophy of Science* 68: 133–161.

French, Steven. 2013. Semi-realism, Sociability and Structure. *Erkenntnis* 78: 1–18.

———. 2014. *The Structure of the World*. Oxford: Oxford University Press.

———. 2015. Between Weasels and Hybrids: What does the Applicability of Mathematics tell us About Ontology? In *Tribute to Patrick Suppes (Tributes* 28*)*, ed. J.-Y. Béziau, D. Krause, and J. R. B. Arenhart, 63–86. London: College Publications.

French, Steven, and Décio Krause. 2006. *Identity in Physics: A Historical, Philosophical, and Formal Analysis*. Oxford: Oxford University Press.

French, Steven, and Kerry McKenzie. 2012. Thinking Outside the Toolbox: Toward a More Productive Engagement Between Metaphysics and Philosophy of Physics. *European Journal of Analytic Philosophy* 8 (1): 42–59.

Glynn, Luke. 2012. Review of Stephen Mumford and Rani Lill Anjum: *Getting Causes from Powers*. Oxford: Oxford University Press. *Mind* 121: 1099–1106.

Heil, John. 2005. Dispositions and Laws of Nature. *Synthese* 144: 343–356.

Horgan, Terence, and Matjaz Potrc. 2008. *Austere Realism*. Cambridge, MA: MIT Press.

Hüttemann, Andreas. 2009. Dispositions in Physics. In *Debating Dispositions*, ed. G. Damschen, R. Schnepf, and K. Stueber, 223–237. Berlin: de Gruyter.

Ladyman, James, and Don Ross. 2007. *Every Thing Must Go: Metaphysics Naturalized*. Oxford: Oxford University Press.

Lange, Marc. 2013. How to Explain the Lorentz Transformations. In *Metaphysics and Science*, ed. M. Tugby and S. Mumford, 73–100. Oxford: Oxford University Press.

Livanios, Vassilios. 2010. Symmetries, Dispositions and Essences. *Philosophical Studies* 148: 295–305.

McKenzie, Kerry. 2013. How Not to Be A Humean Structuralist. In *EPSA 11: Perspectives and Foundational Problems in Philosophy of Science (The European Philosophy of Science Association Proceedings* 2*)*, ed. V. Karakostas and D. Dieks, 307–318. Dordrecht: Springer.

———. Forthcoming. *Between Humeanism and Nomological Essentialism: Towards an Account of Modality in Structuralism.*

McKenzie, Kerry, and Steven French. 2015. Rethinking Outside the Toolbox: Reflecting Again on the Relationship Between Philosophy of Science and Metaphysics. In *Metaphysics in Contemporary Physics (Poznan Studies in the Philosophy of the Sciences and the Humanities* 104*)*, ed. T. Bigaj and C. Wüthrich, 25–54. Amsterdam/New York: Rodopi Brill.

Molnar, George. 1999. Are Dispositions Reducible? *The Philosophical Quarterly* 49: 1–17.

Morrison, Margaret. 2007. Spin: All is Not What it Seems. *Studies in History and Philosophy of Modern Physics* 38: 529–557.

Mumford, Stephen. 1998. *Dispositions*. Oxford: Oxford University Press.

———. 2006. Author's Reply. In: Looking for Laws. Symposium Review on S. Mumford, *Laws in Nature*, Abingdon: Routledge, 2004, by B. Ellis, A. Bird, S. Psillos with a Reply by S. Mumford. *Metascience* 15: 462–469.

———. 2011. Causal Powers and Capacities. In *The Oxford Handbook of Causation*, ed. H. Beebee, P. Menzies, and C. Hitchcock, 265–278. Oxford: Oxford University Press.

Mumford, Stephen, and Rani L. Anjum. 2011. *Getting Causes from Powers*. Oxford: Oxford University Press.

Nolan, Daniel. 2015. Noncausal Dispositions. *Nous* 49: 425–439.

Norton, John D. 2019. The Hole Argument. In *The Stanford Encyclopedia of Philosophy* (Spring 2019 Edition), ed. Edward N. Zalta. https://plato.stanford.edu/archives/spr2019/entries/spacetime-holearg/. Last accessed on 10 May 2019.

Pechlivanidi, Elina, and Stathis Psillos. This volume. What Powers are Not. In *Dispositionalism. Perspectives from Metaphysics and the Philosophy of Science (Synthese Library* 417*)*, ed. A. S. Meincke, 131–149. Cham: Springer.

Pooley, Oliver. 2006. Points, Particles and Structural Realism. In *The Structural Foundations of Quantum Gravity*, ed. D. Rickles, S. French, and J. Saatsi, 83–120. Oxford: Oxford University Press.

Post, Heinz. 1971. Correspondence, Invariance and Heuristics: In Praise of Conservative Induction. *Studies in History and Philosophy of Science Part A* 2 (3): 213–255.

Reutlinger, Alexander. 2013. Metaphysics as a Constraint on Science. Review of John Heil, *The Universe As We Find It*. Oxford: Oxford University Press, 2012. *Metascience* 22, 297–301.

Saatsi, Juha. 2017. Structuralism With and Without Causation. *Synthese* 194: 2255–2271.

Skow, Bradford. 2006. Review of Harvey Brown, *Physical Relativity: Space-time Structure from a Dynamical Perspective*, Oxford: Oxford University Press, 2005. *Notre Dame Philosophical*

Reviews 2006.05.11. https://ndpr.nd.edu/news/physical-relativity-space-time-structure-from-a-dynamical-perspective/.
Teller, Paul. 1986. Relational Holism and Quantum Mechanics. *British Journal for the Philosophy of Science* 37: 71–81.
van Fraassen, Bas C. 1989. *Laws and Symmetry*. Oxford: Clarendon.
———. 1995. 'World' is not a Count Noun. *Nous* 29: 139–157.
Vetter, Barbara. 2009. Review of Alexander Bird: *Nature's Metaphysics: Laws and Properties*. Oxford: Oxford University Press, 2007. *Logical Analysis and History of Philosophy*, 8: 320–328.
Vetter, Babara. 2014. Dispositions Without Conditionals. *Mind* 123: 129–156.
———. 2015. *Potentiality: From Dispositions to Modality*. Oxford: Oxford University Press.
Weinberg, Steven. 2011. Symmetry. A 'Key to Nature's Secrets'. *The New York Review of Books* (October 27, 2011). https://www.nybooks.com/articles/2011/10/27/symmetry-key-natures-secrets/.
Wigner, Eugene P. 2003. Symmetry and Conservation Laws. In *Symmetries in Physics: Philosophical Reflections*, ed. K. Brading and E. Castellani, 23–26. Cambridge: Cambridge University Press.

Chapter 13
Organic Powers

Matthew Tugby

Abstract In this chapter I consider how a realist about powers ought to view the distinction between organisms and non-organisms. This is an underexplored issue but David Oderberg's theory of organic teleology provides a good place for the powers theorists to start. I argue that contemporary realism about powers is conducive to a teleological world view, regardless of whether one accepts the Aristotelian theory of substantial forms that Oderberg favours. According to the theory discussed, organisms are all and only those things that have *self-directed* ('immanent') powers, whereby the possessor of the power is also the subject of the power's effect. Such powers closely resemble what philosophers of biology call *autopoietic* functions. I find this approach promising, but argue that it is not easy to define immanent powers in a way that makes them applicable to all and only organisms. Oderberg and others attempt to draw a sharp distinction between organic and inorganic cases by insisting that only in the organic cases does it make sense to say that the entity flourishes by exercising immanent causation. After exploring possible ways of fleshing out the notion of flourishing, I conclude by considering the possibility that the distinction between organic and inorganic powers is not sharp.

Keywords Powers · Dispositions · Teleology · Organisms · Immanent causation · Purpose · Value

13.1 Introduction: Powers and Organisms

Recently, realism about powers has become a popular metaphysical theory about natural properties. According to this theory, most (if not all) properties are identical with or essentially characterised by causal powers. A distinctive feature of a power is that it is individuated by the manifestation(s) that it is a power for, as when we say

---

M. Tugby (✉)
Department of Philosophy, University of Durham, Durham, UK
e-mail: matthew.tugby@durham.ac.uk

A. S. Meincke (ed.), *Dispositionalism*, Synthese Library 417,
https://doi.org/10.1007/978-3-030-28722-1_13

that solubility is the power *for* dissolving.[1] Realists about powers are strongly opposed to theories on which properties have a primitive and self-contained 'categorical' essence and on which the laws of nature are contingent. The recent rise of the powers theory represents a turn away from the empiricist metaphysical tradition and a return to the idea that causal potentialities are fundamental features of individuals.

There are several metaphysical reasons for finding the powers theory attractive. For example, it has been argued that the rival categorical theory of properties leads to implausible modal consequences, because it implies the metaphysical possibility of properties being able to swap their causal profiles. It has also been argued that with power realism in play, we can provide satisfactory metaphysical analyses of a range of phenomena, such as the laws of nature (e.g. Bird 2007; Mumford 2004), counterfactuals (Jacobs 2010), causation (Mumford and Anjum 2011), natural kinds (Ellis 2001), and modality (Vetter 2015).

Although much research has been undertaken to reveal the core metaphysical implications of the powers theory, less work has been done on how the theory affects our understanding of specific phenomena in the natural sciences. When this latter work has been attempted, it has usually focused on phenomena in physics (e.g. Bird 2007; Nolan 2015). Powers-based analyses of phenomena in the special sciences, such as chemistry and biology, are few and far between. My aim in this article is to help to redress this imbalance by considering a fundamental metaphysical question within the philosophy of biology from the perspective of power realism: what distinguishes organisms from non-organisms? At first glance, this might not seem a particularly interesting question. One might think that the concept of an organism must be commonplace in biology (not to mention everyday discourse) and must thereby be a well understood concept. However, as Nicholson (2014) explains, in the second half of the twentieth century biologists seemed to largely ignore the concept of an organism, shifting their focus to "sub-organismic entities (like genes) on the one hand, and to supra-organismic entities (like populations) on the other", and questions about the nature of organisms were often "dismissed as too metaphysical" (Nicholson 2014, p. 347). Other philosophers of biology have even questioned whether organisms exist at all (see Ruse 1989). Thus, it is far from clear that the concept of an organism is in good standing. Moreover, once we acknowledge that organisms are very diverse—ranging from plants to human beings—it is not at all easy to see what they all have in common. And therefore, it is far from clear what the necessary and sufficient conditions are for being an organism.

If we take the view that powers are fundamental components of reality, then we might expect that the difference between an organism and a non-organism consists in the kind of power that each has. The aim of this chapter is to explore this possibility. The challenge before us, then, is to identify a type of power that all and only organisms share. As we shall see, this challenge is formidable. The structure of the

[1]According to some theories, powers are also individuated by the stimuli which are able to trigger the manifestation of the power (Bird 2007, p. 145).

chapter is as follows. In Sects. 13.2 and 13.3 I introduce the notion of teleological powers and introduce the idea that organisms exhibit a distinctive form of teleology. In Sect. 13.4 I explore the problem of how to characterise organic teleological powers in a way that draws a sharp distinction between organic powers and inorganic powers. In Sect. 13.5 I then consider whether the concept of an organism could still be in good standing even if the boundary between organic and inorganic powers is not sharp. I argue that it could. Although I cannot hope to conclusively settle these debates here, my hope is that the discussion will identify and clarify some of the key issues on which the status of organisms is likely to turn for the powers theorists.

As mentioned above, few powers theorists have considered the nature of organisms using the powers framework. However, I think that the metaphysical work of David Oderberg (2007, 2008) is a natural place to start. Although Oderberg does not always frame his theory in terms of powers,[2] I believe that the teleological aspect of his theory is one that powers theorists should find attractive. According to Oderberg's theory (2008), organisms exhibit a distinctive form of teleology in the sense that they exhibit *immanent* causal processes, which consist in a certain form of *self-directed* behaviour. Within a powers framework, this theory would amount to the view that organisms are distinguished by the possession of immanent powers to act *for their own sake*. I find this immanence hypothesis appealing, and the notion of immanence seems to closely resemble the notion of autopoiesis that is found in the philosophy of biology literature. However, I shall argue that the notion of *immanent* power needs to be spelt out very carefully if it is to delineate organisms and non-organisms. One of the main challenges is that it is difficult to define immanent powers, and the manifestations they give rise to, in a way that precludes them from occurring in inorganic cases of systemic teleology. After considering several ways of characterising organic immanent processes, I shall tentatively suggest that if the theory is to sharply distinguish organisms and non-organisms, it may require an *evaluative* conception of flourishing that has moral import. Some of the challenges facing this approach will then be explored. First, though, we must spell out more of the details of the teleological approach to powers.

13.2 Powers Theory as a Teleological Metaphysics

Teleological metaphysical theories are ones which take certain activities to exhibit 'finality', which is to say they are end-directed in some sense. According to the theistic teleological theories of the medieval period, end-directedness is imposed

---

[2]In his 2008 paper, which is about the difference between organic and inorganic teleology, the terminology of powers is not used at all. However, in more recent work, Oderberg (2017) argues that the contemporary notion of final causation, on which teleology depends, is underpinned by the reality of powers. I shall say more about this later.

externally by the intentions of God. Alternatively, one might claim that end-directedness is *immanent* in the world, grounded in the properties of things. According to Aristotle's version of this view, end-directed behaviours are best explained by the kind essences or 'substantial forms' of things. Oderberg also favours this essentialist view, but it is plausible that the existence of powers alone is sufficient to generate a suitable notion of natural teleology, given that powers are precisely characterised by a directedness to their manifestations. However, to be clear, this teleological notion of end-directedness does not imply the presence of a conscious intention or desire. Rather, this is a naturalistic, non-mental notion of end-directedness. As we shall later in Sect. 13.4, this notion of end-directedness is closely related to the idea of a thing having a natural *function*.

At the heart of any teleological metaphysics is what Aristotelians call 'final' causation. A final cause is precisely a cause which acts or exists for the sake of its natural end or goal. To use Aristotle's biological example of teleological causation, one could say that one of the natural ends of walking is to become healthy (2008, *Physics* II 194b32-195a3). Hence, we might explain (in part) a person's act of walking by reference to the fact that it helps the person to be healthy. But again, not all cases of teleological causation involve the presence of human intentions. According to Aristotelians, non-sentient entities like trees also have powers which exhibit end-directed action, such as an apple tree's power to grow leaves (Cooper 1982, p. 108).[3] What these cases supposedly have in common is that the (teleological) causal explanations involved are forward-looking and this feature generates a contrast with mechanistic causal explanations, which are backward-looking. To explain a state of affairs mechanistically is to explain how that state has arisen from a previous physical structure and the forces governing it. The example of a leaf moving due to the forces imparted on it by a gust of wind embodies the mechanistic paradigm. On the purely mechanistic theory of causation, there are no natural ends, but rather external forces that physical entities blindly obey. Aristotelians do not go as far as to reject mechanistic or 'efficient' forms of explanation, however. Aristotle did not deny that there are mechanistic causes, but his view was that mechanistic regularities are *explained* precisely by the end-directedness or *finality* of things' powers.

In the early-modern period of philosophy (from 1650 onwards), theories of teleological causation were thought by most philosophers and scientists to be of little more than historical interest. The scientific revolution seemed to put mechanistic (or 'efficient') causes at the heart of natural science, making final causes redundant. According to most historical commentators, this early-modern hostility towards natural teleology was generated primarily by scepticism about Aristotelian essentialism and its concept of substantial forms (see e.g. Garrett 1999). The Aristotelian tradition is one which embraces the idea that many things in nature

[3]As Cooper (1982, p. 125) explains, Aristotle was even willing to extend teleological explanations beyond organic cases, as in the case of the frequency of rain in winter and heat in summer. This point will be of importance in the discussion to follow.

have an intrinsic telos, in virtue of their intrinsic natures. However, as acknowledged earlier, within the Aristotelian tradition this reference to 'intrinsic natures' represents a commitment not merely to powers, but to something more: an essentialist thesis concerning substantial forms. For instance, on Aristotle's account the power that an acorn has to become a tree is ultimately explained by the substantial form of being an acorn, where such a form is thought to be irreducible to the attributes that characterise acorns. What is important to note for our purposes, though, is that it is far from clear that a commitment to final causation and natural teleology entails a commitment to the metaphysics of substantial forms. This is a point that is often overlooked, even by powers theorists themselves. A certain theory of powers is arguably sufficient by itself to generate the notion of final causation on which natural teleology relies.[4]

Oderberg (2017) spells out the end-directedness of powers in terms of the notion of specific indifference, which he explains as follows:

> Finality as specific indifference involves two components: (i) a specific range of possible manifestations of a power, and hence a specific range of possible kinds of behaviour by the object having that power; (ii) indifference with respect to the circumstances of manifestation within that range (2017 p. 2394).

Although Oderberg's theory of powers has its nuances, the basic point about powers being directed towards their possible manifestations is shared by most (if not all) powers theorists. As we saw earlier, it is in the nature of a power to tend towards certain manifestations rather than others. Molnar (2003) even argues that the directedness of powers has all of the marks of intentionality (see also Martin and Pfeifer 1986). More recently, Kroll (2017) argues that disposition concepts should be *defined* in terms of end-directedness, leaving us with what he regards as a teleological analysis. It would hardly be surprising, then, if the powers theory led us naturally towards a teleological account of natural phenomena such as organisms.

If the points above are correct, then it seems that Oderberg's own preference for a theory of substantial forms is dispensable as far as the theory of final causation is concerned, providing we are realists about powers. This is a point that Oderberg himself acknowledges when he says that "one might hold both that particular kinds of directedness were real phenomena of particular kinds of object and that they were merely accidental to those kinds" (2017, p. 2398).[5] I think it is all to the good that the powers theory can accommodate finality in nature without taking a stance on the controversial issue of whether there is a distinct ontological category of substance kind essences. This allows the powers theorists to evade most early-modern critiques

[4]Oderberg is clear that he thinks organic teleology can be understood in terms of final causation, even though he prefers the terminology of immanent causation (2008, p. 263).

[5]Similarly, in his 2008 paper Oderberg mentions in passing that his account of the distinction between organic and inorganic teleology does not presuppose Aristotelian essentialism (2008, p. 264). For example, all that his account of inorganic teleology requires is "the thought that for something that is recognizably the rock cycle or the water cycle on Earth to occur, certain kinds of thing have to play certain kinds of role, and certain kinds of processes have to take place" (2008, p. 272).

of Aristotelianism and the difficult questions that substance kind essentialism invites. I do not have the space to discuss these issues in any detail but one obvious difficulty concerns what the essence of a substantial form is like if it is irreducible to, and something-over-and-above, the attributes that characterise it. Moreover, how can a substantial form's essence, which is supposedly simple, explain something complex like an object's causal profile? Of course, if such difficulties are insurmountable, this would not mean that powers theorists are not entitled to speak of natural kinds. Indeed, the property of being an organism looks precisely like a high-level natural kind. But as the powers theorist Mumford explains, it is perfectly coherent for an anti-essentialist powers theorist to talk seriously about natural kinds. In Mumford's view, all that talk of kinds requires is that "each kind-member instantiates the appropriate properties" (2005, p. 420).

A further feature of a powers-based teleological approach to organisms that is worth mentioning is that it can remain silent about how organisms come to *have* the powers that characterise them. I think this neutrality is to be welcomed. We know through the work of Darwinians that organic species like ours acquire their characteristics through natural selection. But it would seem too hasty to build a requirement into our metaphysical theory of organisms that organic powers *must* be naturally selected. This would rule out the metaphysical possibility of organisms whose powers have not evolved through natural selection. Ruling this out seems too strong. Even if, in worlds like ours, all life arises through natural selection, distant possible worlds seem conceivable in which beings like us have a different causal history.[6] Moreover, it might even be the case in the actual world that there will be entities that are alive but which have been designed rather than evolved. For example, if powers are the building blocks of reality, then the functionalist theory of mind (or something like it) has some plausibility, which means that powers theorists should be not be too hasty in ruling out the metaphysical possibility of artificially intelligent agents that are alive.[7]

Now that the basic motivations for a powers-based natural teleology have been sketched, let us consider in more detail what Oderberg says about the sort of teleology exhibited by organisms, and begin to clarify the notion of *immanence* on which it relies.

---

[6]For instance, perhaps there is a possible world in which an organism is formed through a purely chance encounter of different fundamental particles.

[7]A more pragmatic reason for avoiding debates about natural selection in the current context is that it is far from clear how Darwin's theory sits with respect to teleology. Opinions on this issue differ widely (see e.g. Ariew 2007; Lennox 1993). For current purposes, we shall simply understand teleology in terms of the end-directedness of the powers of individual organisms. Whether or not the mechanisms of natural selection at the level of species can be regarded as involving a form of teleology is a question we need not address here.

13.3 Organisms and Immanent Powers

What, then, does a promising powers-based teleological account of organisms look like? As mentioned earlier, Oderberg's view is that all and only organisms engage in *immanent* causation:

> Speaking now in causal terms, living things, unlike non-living things, exercise immanent causation: this is a kind of causation that begins with the agent and terminates in the agent for the sake of the agent (2008, p. 261).

Although Oderberg does not always spell out his theory using the terminology of powers, we can easily construe the theory as proposing that all and only organisms have *immanent powers*. The causation that immanent powers generate is contrasted with what Oderberg calls "transient" causation (2008, p. 262), which occurs when a cause and its effect involve distinct entities. The paradigmatic cases of immanent causation are those in which an organism "[. . .] acts so as to produce, conserve and repair its proper functioning as the kind of thing it is" (2008, p. 261).[8] Oderberg does not deny that some organic immanent processes involve transient interactions between an organism's parts, such as the causal exchanges between different digestive organs. Digestive processes may also have transient effects which go beyond the organism, as when an organism excretes. Hence, it would not be correct to say that the powers of organisms are *purely* immanent. Nonetheless, certain aspects of the digestive process are immanent to (say) a human being, in the sense that they take place within, and for the benefit of, that same human being. The basic idea behind an immanent powers criterion, then, is that it is only organisms which *act for themselves*. Feser calls this a Scholastic view of organisms: "For Scholastic writers, a capacity for this sort of 'immanent causation' (to use the Scholastic jargon) *just is* what makes something a living thing" (2010, pp. 149–150).

In order to assess the plausibility of this theory, we must first consider the sufficiency of the immanent powers proposal. Is it really the case that all entities which have the capacity for immanent activity are organisms? As immanent powers have been characterised thus far, it is far from clear that they are distinctively organic. If the account is to be plausible, we need to say more about what is meant by 'immanent causation'. Care is needed here because, as shall now see, the terminology of immanent causation is often used in a different way by other metaphysicians. What this shows is that immanent causation in Oderberg's sense has to be understood in a specific way.

One of the features of immanent powers that we have focused on thus far is that their causes and effects concern *one and the same entity*. However, surely the biological realm is not the only realm in which immanent causation in this broad sense takes place. The immanent/transient (or 'transeunt') distinction is well known in the general causation literature, as Armstrong's work illustrates. Armstrong (1997,

---

[8] Again, talk of kinds here need not be interpreted in strong essentialist terms, even though that is the view that Oderberg prefers.

p. 73) says of immanent causation that it is a "remaining within" causality, whereas "transeunt" causation is a "going across" causality. Importantly, Armstrong goes on to offer examples of immanent causation in this broad sense which do not involve organisms. One example concerns the decaying powers of a radioactive atom. Such cases involve radioactive emissions that do not appear to be triggered by any events external to the atom (1997, p. 74). Another putative example of immanence offered by Armstrong concerns the persistence of an object through time. Armstrong claims that if one accepts a perdurantist temporal parts account of persistence, then one requires a principle of unity "by which non-overlapping temporal parts of the one particular are welded together to constitute the single thing that exists through time" (1997, p. 74). Armstrong concludes that in the face of this challenge, the best thing to say is that persistence is always a matter of an entity immanently causing itself (including its properties) to exist from one moment to the next.[9]

Now, it must be acknowledged that Armstrong's examples of immanent causation are controversial. Cases in fundamental physics are notoriously difficult to interpret, and one might follow Cartwright (1989, p. 109) and Ellis (2001, p. 129) in thinking that spontaneous radioactive emissions are uncaused events. However, it seems to me that there are examples of inorganic immanent causation in Armstrong's broad sense that are less controversial than the ones he appeals to. For example, in a critical discussion about the transference theory of causation, Dowe (1995, p. 367) refers to the case of a space ship's inertia being the cause of its continuing motion. In this case, there is no transfer of energy from one object to another, and so this example does not appear to involve any transient causation. Yet, it seems plausible that causation is taking place: unlike Armstrong's examples of static persistence, something *is* clearly happening in the inertia case and the movement involved obeys well known dynamic laws.

In summary, it seems that if organisms are to be characterised as all and only those things which possess immanent powers then we will have to provide a more specific notion of immanent causation than that which is ordinarily employed in the causation literature. In the next section, we shall examine the specific features that Oderberg attributes to immanent causation in the organic sense. The main conclusion of that section will be that Oderberg's concept of organic flourishing is crucial, and two possible ways of fleshing out this concept are scrutinised—one of which is normative and one of which is evaluative.

[9]On questions regarding persistence and dispositionalism see also Williams (Chap. 6, this volume) and Meincke (Chap. 7, this volume).

13.4 Organic Immanent Behaviour: Possible Interpretations

13.4.1 Immanent Powers as Autopoietic Functions

If the observations above are correct, then cases of immanent causation in Oderberg's sense, which are supposed to be the mark of the organic, represent a mere proper subset of what many metaphysicians think of as examples of immanent causation. What, then, distinguishes organic immanent powers from other powers? As mentioned earlier, Oderberg's idea is that immanent causation is his sense is "a kind of causation that begins *with* the agent and terminates *in* the agent for the sake *of* the agent" (2008, p. 261). Given that the cases discussed in the previous section are ones in which the causation begins with and terminates in one and the same entity, it has to be the notion of *acting for one's own sake* which distinguishes Oderberg's notion of immanent causation from that discussed by people like Armstrong. What, though, does it mean to act for one's own sake? Oderberg's answer is that a living thing acts for itself in the sense that it "acts so as to produce, conserve and repair its proper functioning as the kind of thing it is" (2008, p. 261). At the heart of this account, then, is the notion that immanent causation in the organic sense has a *self-maintaining* role, in the sense that it ensures the organism survives and continues to behave in ways that are normal for it as the kind of thing it is. In line with our comments earlier, although Oderberg favours an essentialist view of kinds, one could interpret this reference to kinds in a milder way that does not involve essentialist commitments.

Although Oderberg does not use the following terminology, it seems that this initial characterisation of immanence resembles what philosophers of biologists call *autopoiesis* (following Maturana and Varela 1980). As Nolt (2009) explains, autopoietic functions are those which promote the survivability of an entity. More precisely, the autopoietic functions of organisms

> establish, maintain or enhance their survivability—functions such as capturing sunlight or prey, resisting disease, obtaining water from the environment, respirating, healing injuries, eliminating wastes, and so on. (Nolt 2009, p. 149)[10]

In short, both Oderberg's notion of immanent processes and the notion of autopoietic processes concern behaviours which allow an organism to maintain *itself* in a systematic way. However, it would be a mistake to think that *all* the functions of

---

[10]As Anne Sophie Meincke has pointed out to me, the resemblance may not be perfect because Maturana and Varela's understanding of autopoietic functioning seems to be more specific than that employed by Nolt. For Maturana and Varela (1980), it is not merely that autopoietic functions promote the survival of the organism but rather the functions themselves are continually realised and regenerated through the structural organisation of the organism. On this definition, autopoiesis can only be said to occur when certain processes of production of material components are in place—processes which ensure the stability of the organism through time. For a fuller discussion of this concept of autopoiesis and some of its ontological implications, see Meincke (2019).

an organism are autopoietic or immanent in Oderberg's sense. For example, Nolt (2009, p. 261) contrasts autopoietic functions with "exopoietic" ones, an example of the latter being the power to reproduce. Although the power to reproduce allows an organism's species to survive, it is not autopoietic because reproduction may not aid the survival of the organism that reproduces. For example, successful reproduction comes at the cost of death for some organisms, as in the case of some female octopi, which starve while protecting and caring for their eggs.

Is it plausible, then, that the capacity for immanent or autopoietic behaviour is the mark of an organism? Oderberg's discussion of immanence suggests that it is, but there are two worries about this criterion that need to be addressed. First, it is questionable whether the immanent powers characterised thus far can be found only in organic cases. Second, even if Oderberg's account of organisms is extensionally correct, it is questionable whether it can be articulated in a non-circular way, a way which does not presuppose the concept of being an organism.

Surprisingly, my reasons for thinking that immanent powers (in Oderberg's sense) might not be restricted to organic realms are inspired by putative examples of inorganic systemic teleology that Oderberg himself discusses. Oderberg wants to maintain that inorganic systemic teleology is in some way second-rate, in the sense that it does not involve *bona fide* immanent behaviour. But as we shall see, this move is not easy to make.

Oderberg's examples are those of the water cycle and the rock cycle (2008, pp. 266–8). These are cases in which there is ordered, systematic behaviour that leads to the stable recurrence of certain processes (2008, p. 271). In the case of the water cycle, there is a three stage cyclic process of condensation, followed by precipitation, and finally evaporation, which then leads back to condensation and the regeneration of the cycle.[11] Because each step in the cycle is dependent on others in a specific way, Oderberg concludes that the steps in the cycles have *role specific* functions (2008, p. 272), which ensure the stability and maintenance of the entire system. He defines this notion of function as follows:

> x performs an inorganic function with respect to y = def. x is inorganic and y is inorganic and x contributes causally to some entity, event or process in y and y is a stable, systematic process. (2008, p. 273)

Now, one might react to this functional analysis of the water cycle in one of two ways. First, one might agree that talk of performing a function is appropriate in the water cycle case and accept that the water cycle exhibits teleology in some sense. This is the route that Oderberg takes. Alternatively, one might resist the notion of inorganic teleology and insist that talk of performing a function is only applicable in organic cases. As Oderberg (2008, p. 270) points out, this is the route that most teleologists would take (see e.g. Bedau 1992). However, as acknowledged earlier, I do not think powers theorists should question this part of Oderberg's analysis. For current purposes, the important question is whether, if we accept a teleological

[11]This is a simplified explanation, but it will suffice for our purposes. Clearly, each step described above is constituted by many complex sub-processes.

analysis of the water cycle, a robust distinction can be maintained between organic and inorganic powers.

The problem is that the water cycle seems to have many of the hallmarks of immanent or autopoietic powers characterised earlier. Cases of immanent causation are initially characterised by Oderberg as those which "produce, conserve and repair" the other proper functions of a thing (2008, p. 261). Elsewhere Oderberg (2008, p. 263) also associates immanent causation with processes which ensure the *continued existence* of a thing. The example of the water cycle seems consistent with these features. Crucially, the notion of systemic stability is used by Oderberg to explain why the water cycle is teleological. For instance, Oderberg writes that "I contend that the mere stability and recurrence of certain processes such as the rock and water cycles license teleological talk in terms of functions and roles going beyond mere causation" (2008, p. 271). The important point to note here is that the stability and recurrence that Oderberg speaks of ensures the continued existence/ survival of the relevant systems and their proper functioning. Hence, it is far from obvious that this kind of teleology differs significantly from the immanent or autopoietic activity characterised earlier.[12] Moreover, Oderberg's notion of role-specific functions in the water cycle arguably allows us to make sense of the idea of the elements of the system acting *for the sake* of the whole of which they are parts: the role-specific functions are defined in terms of their contributions to other stages of the cycle, all of which mutually sustain the system.[13]

Fortunately, Oderberg does have more to say about what distinguishes organic teleology from the sort of systemic teleology that (we are assuming) is present in the water and rock cycles. Oderberg's response to the sort of challenge just outlined is that the notion of function that is applicable in inorganic cases "[. . .] is divorced from the idea of any intrinsic purpose, immanence or principle of flourishing" (2008, p. 269). What becomes clear, then, is that immanent behaviour in Oderberg's sense is laden with intrinsic purpose. For Oderberg, this is to say that organisms act for their own sake in the sense that they promote their own flourishing.[14] It is arguable

[12]Perhaps one could place significant weight on Oderberg's notion of repairing as way of distinguishing inorganic and organic teleology. However, it is far from obvious that the notion of repairing is inappropriate in the case of a water cycle, if repairing means returning oneself to normal functioning following a disruption. For example, the natural rhythm of a water cycle is disrupted during periods of extreme drought, which typically occurs when the atmospheric conditions of the cycle block the upward forces necessary for moisture to precipitate. What is important to note, though, is that the water cycle is disposed to return to its natural rhythm as soon as the obstructions are no longer there.

[13]For example, Oderberg notes that "If water is to be precipitated, then condensation or something very like it has to take place. Evaporation of surface water is going to produce clouds or something very like them" (2008, p. 272).

[14]Another possible difference that Oderberg alludes to is that inorganic systems can be instrumental causes for both inorganic and organic entities whereas organic systems cannot be instrumental causes for inorganic entities (2008, p. 266). However, we shall not discuss this suggestion here, because even Oderberg acknowledges later in his paper that there may be exceptions to this rule. For example, "some organic things respire, and respiration is part of the water cycle. So to that extent

that by appealing to the notion of flourishing, Oderberg's notion of immanence moves away from the notion of autopoiesis, which relies more on the notion of systematic regeneration than flourishing, at least in its orthodox formulations. If this is right, then we could accept that the water cycle exhibits something like autopoietic powers while insisting that those features are not immanent in Oderberg's sense.[15] According to this proposal, which we shall explore in the next section, organic powers remain distinctive due to their immanence.

13.4.2 Organic Powers, Intrinsic Purpose, and Flourishing

Let us consider in more detail how one could defend the claim that the notions of intrinsic purpose and flourishing are inapplicable in cases of inorganic teleology. It should be noted that when applying the concept of function to the examples of the rock and water cycles, Oderberg does acknowledge that there is a way of reading

> X has a function of doing Y' that makes it equivalent to saying either that X has a purpose for which it does Y, or that the thing in respect of which Y is done has a purpose for which X does Y in respect of it. (2008, p. 270)

However, Oderberg asserts that if this is what someone meant by 'function', then he would retract his claim that the rock and water cycles have functions. This again illustrates his idea that the concept of inorganic function is "divorced from the idea of any intrinsic purpose, immanence or principle of flourishing" (2008, p. 269). I find this move difficult to justify, however. The main problem is that Oderberg must already accept that there are different kinds of purpose because he urges that purposes do not only arise in cases where there are intentions or desires. For instance, the fact that the pumping of the heart helps to fulfil an organism's purpose to stay alive "[. . .] does not imply conscious activity, or any idea to the effect that the organism *tries* or *seeks* to keep itself alive by using the heart as a means" (Oderberg 2008, p. 263). Indeed, some organisms, such as plants, are not capable of having thoughts at all. This is why intentions cannot be necessary for purposeful action, because if they were then even plants would not count as organisms on Oderberg's account. My point, then, is that Oderberg must have a liberal account of what counts as purposeful action, as any biological teleologist should. Even very basic organisms such as bacteria have intrinsic purposes on Oderberg's account. However, once this is acknowledged, it is far from obvious why some inorganic systems cannot also be said to display purposeful action in some non-intentional sense. I admit this is not a conclusive argument, but what it does suggest is that more needs to be said about the

---

one could say that organic things and processes can instrumentally serve inorganic ones" (2008, p. 276).

[15]This would be consistent with Nolt's view about autopoiesis, which is that some non-biological artefacts such as robots might come to exhibit it (2009, p. 143).

notion of 'intrinsic purpose' and 'principle of flourishing' that is in play before we can draw a firm verdict.

When discussing a slightly different issue, Oderberg makes some comments which could potentially be used to assuage the problem just outlined. In one place he considers the worry that functional talk cannot be applied in inorganic cases because talk of functions implies purpose. Here, Oderberg's objector agrees that there is no purpose in inorganic cases but, unlike Oderberg, thinks for that reason that such cases cannot be teleological. Oderberg's reply is that talk of functions does not commit one to the existence of purposes because in his description of the water and rock cycles, he was able to describe their respective functions without ascribing intrinsic purposes to them (2008, p. 274). Perhaps this same comment could also help with the worry raised toward the end of the previous paragraph: the argument would be that there are no purposes in inorganic cases because their functions can be described adequately without the ascription of purposes. I have a couple of related concerns about this kind of response, however. First, surely it is possible to describe the functions of very simple organisms, or even complex organisms like plants, without ascribing intrinsic purposes to them. But it would clearly be a mistake for Oderberg to conclude from this fact that these organisms lack purpose, for then they would not count as exhibiting organic teleology by his own lights. Second, and relatedly, Oderberg is open to the charge that by describing the water and rock cycles in the way he does—by omitting talk of purpose—he has merely provided a partial characterisation of the relevant functions. Compare: Aristotelians think we can (and often do) describe natural processes in terms of mechanistic or 'efficient' causation, without ever mentioning final causation. However, this does not show that final causation does not exist. For the Aristotelians, the full metaphysical story is that final causation is an underlying precondition for efficient causation.[16]

However, even if my responses here are strong, perhaps Oderberg would emphasise that his specific notion of intrinsic purpose is tied to the notion of flourishing. For Oderberg, the notions of intrinsic purpose and flourishing are two sides of the same coin: the flourishing of an organism requires that it "has an intrinsic telos, a principle of natural fulfilment, such that it characteristically behaves in such a way as to achieve or seek to achieve that fulfilment" (2008, p. 265). This strategy involves insisting that purposeful action is only ascribable in cases in which it is intelligible to say that a course of action is *good for* the object. On this account, the existence of purpose entails the existence of natural goods which the object pursues.[17] Oderberg

[16]Indeed, Oderberg (2017, p. 2396) endorses this point in a recent paper: "Final causes are the precondition of the very possibility of any efficient causality." Nonetheless, properties of finality may not be 'the direct object' of scientific investigation (ibid. p. 2400).

[17]It is worth noting another possible way of justifying the idea that all and only organisms have natural purposes, which is to argue that all and only organisms have awareness and that awareness entails purpose. This is not a strategy that is available to Oderberg, however. Oderberg is happy to say that simple non-vegetative organisms have a primitive awareness of their surroundings (2007, p. 192), but he does not extend this claim to plants (2007, p. 187) because they only have motor organs (as opposed to sense organs). Hence, if the presence of primitive awareness were the

is clear that, in this sense, all organisms have powers to flourish, as when he says that "[b]acteria seek to flourish every bit as much as human beings" (2008, p. 265). In contrast, Oderberg insists it is absurd to think that inorganic entities flourish: "Rocks do not flourish; there is nothing that is good for evaporation [. . .]" (2008, p. 274).

What, though, is meant by 'flourish' and 'good for' This is not always clear. Moreover, it may be that the notion of organic functioning is conceptually prior to the notion of flourishing, so that the former cannot be reductively analysed in terms of the latter. Indeed, Oderberg seems to gesture in this direction when he writes:

> If there were such a thing as inorganic teleology, what differences would we expect to see between it and the organic case? They should be derived from our prior understanding of what is characteristic of the living and the non-living. (2008, p. 264)

Since Oderberg's notion of organic teleology is inextricably linked with his notion of flourishing, this quote suggests that we might not be able to understand the notion of flourishing independently of what it is to be alive.

Where does all this leave us? I think the foregoing observations indicate that we face a dilemma. If we cannot understand the notions of purpose/immanence/flourishing independently of what it is to be an organism, then although it will be true by definition that all and only organisms have immanent powers, we will not be able to give a non-circular explanation of what it is to be an organism *in terms of* those powers.[18] On the other hand, if these notions are not conceptually tied to the notion of being organic, it is difficult to see why it is absurd to say that the water cycle is flourishing when the role-specific functions of its parts are performing successfully. Hence, if we are to accept that inorganic teleological systems do not flourish or act for the good of themselves, it would be helpful to have an informative and non-circular account of flourishing that explains why this is the case. In the next section I shall consider a couple of possible options in this direction, the most plausible of which ties the notion of flourishing to the notion of inherent value.

13.4.3 Autopoietic Powers, Normativity, and Value

One obvious route to take is to explicate the notion of flourishing in terms of a concept of natural normativity. Often when we ascribe goodness to an action, we intend our claim to have normative force. For instance, when we tell children that eating greens is good for them, we intend to convey that for their own good they *ought* to eat greens. With this in mind, one might try to define organic immanent

---

underlying criterion for being an organism, then plants would not qualify as organisms on Oderberg's theory.

[18] Again, Oderberg might not be unhappy with this result, because the primary aim of his 2008 paper is to show that there are inorganic as well as organic cases of systemic teleology. Nonetheless, in the current context this is not a welcome consequence, because our hope was that an appropriate notion of power could help to provide a non-circular analysis of what it is to be an organism.

powers as those which their possessors ought to manifest for themselves. The thought would then be that all processes which are inorganic fail to have this normative aspect. For instance, we might argue that although the water cycle exhibits teleological features, in virtue of the powers involved, there is no sense in which the water cycle *ought* to evaporate (or precipitate) water.

There seems to be something right about the normative criterion, but if it is to be defensible we need a metaphysical account of how a natural organic process can exhibit normative features.

The best attempt that I know of to generate a naturalistic notion of normativity within the powers framework is found in the work of E. J. Lowe (1980, 1982) through his theory of laws.[19] Lowe's basic idea is that if laws reside in the inherent powers of things, then there is a sense in which an entity ought to behave in one way rather than another in a given environment. On this account, law statements are not merely descriptive, as they are on Humean regularity theories. Rather, laws are generated by the essences of things, which play a *regulative* role. The essential powers of things dictate how they ought to behave, if they are normal exemplars of their kind. Consider, for instance, the claim that bees are disposed to fly. This seems like a law statement in some sense, and yet it is not equivalent to the regularity statement that all actual bees fly. Clearly, a bee could live without flying. For instance, a bee might get stuck in a hive, or it might decide to crawl up to the flowers on which it feeds rather than fly between them. However, according to Lowe's theory, it remains the case that the bee *ought* to fly, in virtue of its nature as a bee. In explaining the connections between disposition ascriptions and normativity, Lowe draws an analogy with legal laws. Legal laws have a prescriptive rather than descriptive force. They do not describe how people actually behave but rather dictate how they should behave.

Can the Lowean approach help to underpin an account of organisms that is based on the normativity of certain powers? Unfortunately, it is far from clear that this approach is workable. The first salient point is that Lowe's normative theory of laws is underpinned by his four-category ontology, which contains the category of

[19]Oderberg's own theory of laws (2010) is not dissimilar to Lowe's, but as far as I can tell his theory of organic flourishing does not rely on it, at least not explicitly. I should also note that there is a lively debate in the philosophy of biology about whether a naturalised notion of normativity is applicable in biological contexts. For example, Barandiaran and Moreno (2008) argue that organisms exhibit intrinsic normative functioning on the basis of their autonomous adaptive organisation, while Barham (2012) argues that organisms exhibit normative agency on the basis that they act to preserve their own existence. Since space is limited I shall focus only on Lowe's metaphysical account here. And for the purposes of this chapter, we need not deny that there is normative organisation and behaviour in the biological realm. I note, however, that for the sorts of reasons discussed below and also by Walsh (2008, pp. 120–121), there are reasons for doubting that the normative aspect of biological organisation is what distinguishes organisms from non-organisms normativity will plausibly occur in the inorganic cases of systemic teleology discussed earlier. And if the normative claim about certain biological cases is meant to capture something stronger, such as the presence of intentional action, then we must accept that some organisms will not exhibit the relevant normative behaviour (Walsh 2008, p. 121).

substantial forms or kinds (Lowe 2009). On Lowe's theory, it is the essences of substantial kinds, such as 'beehood', which ultimately grounds facts about how exemplars of that kind ought to behave. As we saw earlier in Sect. 13.2, this is a considerable metaphysical commitment that goes beyond power realism, and it is one that many metaphysicians may not be prepared to make. But pushing that detail aside, what is more problematic is that if Lowe's theory of natural normativity is successful, it provides us with a *general* account of laws. Since not all laws concern organic entities, the Lowean approach entails that there are also normative facts about how members of inorganic kinds ought to behave. For example, Lowe's theory implies that a lump of salt ought to manifest its power to dissolve when placed in water, in virtue of being a lump of salt. In short, then, Lowe's normative theory of powers and laws would not allow us to draw a distinction between organic and inorganic powers where natural normativity is concerned.

Fortunately, even if we deny that natural normativity is the defining feature of organic power, surely there might be other ways to maintain that there is an important difference between organic and inorganic activity. One route is to adopt a strong evaluative criterion for organic powers, which would say (roughly) that the difference between organic and inorganic powers is that only the manifestations of the former are of *inherent value* to their possessors.[20] Within the power realism framework, the idea would be that all and only organisms have certain powers whose activation realises the second-order property of *being valuable* for the possessor of those powers, where such a property is regarded as being human-independent and irreducible. It seems that, first and foremost, it will be the manifestations of autopoietic powers that are of value to their possessors, given that they ensure their survival. However, such an account can leave it open as to whether other kinds of power manifestation are also of inherent value for organisms. As we saw earlier, exercises of transient causal relations often arise from autopoietic processes (as when excretion occurs through digestion), which suggests that some transient relations might also be of value in a derivative sense.

An evaluative account of organic behaviour is of course consistent with a normative understanding of organic behaviour. Indeed, on Bedau's (1992) account, a goal is only something a subject ought to attain if that goal is good for them. However, Bedau's own understanding of goal value seems too liberal to distinguish organic teleology from the cases of inorganic teleology that we accepted earlier. In particular, for Bedau teleological goodness need not consist in moral goodness. Rather, organic goals are good in some cases merely insofar as they are "useful or beneficial" (1992, p. 791). But if, as discussed above, it is intelligible that inorganic teleology exhibits goal-directed behaviour, then there is no obvious reason why the concept of usefulness cannot be applied in those inorganic cases too. The same goes

[20]For the sake of brevity, I will sometimes speak of the organic autopoietic powers as being of value to their possessors rather than the manifestations of those powers. However, strictly speaking perhaps we should say that the powers themselves are valuable in a derivative sense, insofar as they are responsible for bringing about inherently valuable manifestations which maintain the organism.

for Ayala's evaluative account of teleological goals, which takes such goals to be good merely in the sense of having utility, where utility means contributing "to the reproductive efficiency of the organism itself" (1970, p. 13).

With the preceding points in mind, it is natural to consider a stronger theory of value in organic cases, for example one which says that the value realised by organic powers makes moral demands upon those capable of moral deliberation. This proposal is supported by the fact that humans capable of moral deliberation normally feel morally obliged not to prevent (say) a horse from exercising its power to eat grass, suggesting that we tend to regard the eating of grass to be of value to the horse—even if the horse is not consciously aware of this. The same cannot be said of inorganic entities such as rocks, however.

Before considering one of the challenges facing this strong evaluative theory, some further clarifications are in order. If this evaluative criterion is to provide a plausible account of the organic/inorganic distinction, then we cannot accept that the value claims about organic powers (and only organic powers) are the result of humans projecting their own values onto the world. If we accepted this, then the organic/inorganic distinction would itself be human-dependent, which is implausible. This is not to say that we do not also have perfectly good human-dependent reasons for valuing plants and animals. Clearly, we need plants to sustain oxygen levels for our own survival. However, if the evaluative criterion is to provide an *objective* account of the distinction between organic and inorganic entities, the distinction should not rest on values which depend on contingent facts about what is instrumentally good for humans, not least because inorganic systems such as the water cycle might also be of instrumental value to beings like us.

What these considerations show is that if the evaluative criterion of organisms is to succeed, we must commit to a realist view about value and defend the claim that it is only organic powers whose manifestations are *inherently* valuable for their subjects, unlike inorganic powers. To be clear, these value properties would be irreducible and so not themselves susceptible to a reductive analysis in biological terms. For this reason, the theory would not leave us with a circular account of what it is to be organic. Since, on this picture, the value realised by certain powers will be ontologically primitive, the theory does not depend on a prior grasp of what it is to be organic. Rather, our grasp of such properties will depend only on our having the right sorts of moral sensibilities. We shall now explore this evaluative account of flourishing in more detail.

13.4.4 The Evaluative Account of Flourishing: Costs and Challenges

Perhaps the main theoretical cost of the evaluative criterion is that the notion of inherent value will strike some as obscure, because it has to be taken as a primitive notion. However, this seems like an inevitable feature of the evaluative theory of flourishing that I am considering, precisely because of the difficulty of distinguishing organisms and non-organisms in any other terms. As we have seen, it is far from obvious that the notion of autopoiesis as defined by Nolt can be used to generate the distinction between organic and inorganic activity. Nolt for one would agree, since he sees no reason why autopoiesis could not be exhibited by artefacts such as machines (2009, p. 143). Moreover, it is difficult to see how the notion of inherent value in play can be explained in terms of other properties, such as the property of being sentient. It is true that when people feel morally obligated in various ways towards other organisms, it is usually because those organisms are capable of suffering[21] when their natural ends are frustrated, and this capability for suffering plausibly arises from their being sentient. So, one might be inclined to think that the inherent value of some powers is ultimately grounded in sentience. However, the problem in the current context is that a sentience-based account of inherent value would not go far enough. Unless we adopt a radical form of panpsychism, it is not at all plausible that all organisms are sentient, at least not in the sense that they can be said to experience suffering. Nor would it help to insist that all organisms (sentient or not) have a primitive awareness of their environment and to explain the presence of inherent value in terms of the presence of primitive awareness. The problem is that if we say that simple organic entities such as single cells have primitive awareness, then we have watered down the notion of awareness so much that it would arguably occur in inorganic cases, as in the case of machines which are sensitive to changes in their environment. In that case, the notion of inherent value could not be used to distinguish cases of organic and inorganic powers. It seems plausible, then, that if the evaluative criterion is to be successful, we will need to accept a full-blooded value realism on which it is an irreducible fact that the manifestation of a certain power (and, derivatively, the power itself) is inherently valuable for the possessor of that power in a morally relevant sense.

Assuming that we are happy to accept that inherent value is fundamental, what challenges does the evaluative theory of organic power face? One puzzle concerns the extent to which different autopoietic powers in the organic realm have the same amount of inherent value. On the one hand, it seems implausible to suppose that the autopoietic processes of organisms like single cells or plants have the same moral significance as those of human beings. But on the other hand, some philosophers are reluctant to accept that the activities of different organisms have varying moral status. This last point is of particular concern to some environmental ethicists. In line with the theory we have sketched above, both Taylor and Regan maintain that certain courses of events are valuable *for* organisms in some objective, morally relevant sense (e.g. Taylor 1986; Regan 2004). However, Regan is reluctant to deny that the flourishing of all organisms has equal moral worth because if we maintain that the value realised by organic powers comes in degrees, this might pave the way for "ethically unacceptable forms of subjugation" with respect to animals and some human beings (2004, p. 247). How, then, can this dilemma be resolved?

[21]See Singer 1990 on the connection between the capability of suffering and moral relevance.

Nolt's response to this issue is to question the claim that all kinds of organic autopoietic behaviours are of equal value to their possessors. Nolt's worry (2009, p. 146) about Regan's equality claim is that it implicitly appeals to an anthropogenic (i.e., human dependent) standard of justice. Nolt's argument is that the equality claim is not plausible if we base it purely on the intrinsic features of organisms. For example, it seems clear that not all autopoietic processes carry the same degree of self-concern when we compare, say, a human being with a plant. Nolt concludes, therefore, that in so far as (say) claims about animal equality are plausible, they must rest on an anthropogenic theory of justice. If Nolt is right about this, then it seems that claims about the value of certain power manifestations reflect facts about humans as much as they do the world. This, in turn, would make our evaluative criterion for organisms implausible, for what is needed is a human-independent notion of organic value, if the organic/inorganic distinction is to be an objective one.

How, then, can a value theorist deal with the question of relative moral worth? I do not have a detailed solution to offer here because settling the issue would take us deep into bioethics and the metaphysics of value. It would be unrealistic to attempt that work here, but we can nonetheless try to advance the debate by identifying three strategies that could be employed in order to defend the view that all (and only) organic autopoietic powers are of inherent value to their possessors. My hope is that this will provide a platform for future philosophical research on the topic.

One strategy is to agree with the claim of Regan and others that the autopoietic powers of different organisms carry equal inherent value for them and to insist that this judgement is not based on anthropogenic standards of justice. Nolt fails to find any biological properties which can plausibly ground the equality claim, given that organisms are biologically so diverse. However, if we accept the strong form of value realism that we have outlined, we can insist that value properties are irreducible and cannot be explained *in terms* of biological concepts. Rather, we will think of inherent value as a fundamental second-order property that is realised for an organism when it acts for itself, using its autopoietic powers (and perhaps certain others). Biology by itself will not reveal these second-order evaluative properties. Rather, such properties would only come into view through the lens of moral deliberation. According to this theory of value, value properties are akin to secondary qualities, realistically construed (McDowell 1985). Just as eyes are necessary for bringing the (real) colours of things into view, a certain moral sensitivity is needed to bring the irreducible and inherent value of certain powers into view. Note also that, importantly, the powers theory seems particularly conducive to this approach given that secondary qualities are typically thought to be dispositional in nature.[22]

A second and alternative strategy is to accept that all organic autopoietic powers carry inherent value for their possessors, but to reject the claim that such powers exhibit *equal* inherent value. In order to avoid Regan's ethical unease about such a proposal, one would have to argue that it does not follow that unacceptable forms of

[22]Most powers theorists will of course insist that primary qualities are also conferrers of dispositions.

subjugation are morally permissible. The first part of this response could be supported by examples that Nolt himself discusses. When considering cases in which human bodies are in a permanent vegetative (i.e., non-sentient) state, Nolt (2006, p. 363) admits to no longer feeling confident that the organism demands the same level of moral reverence as a fully functioning human. If this intuition is correct, does it follow that unacceptable forms of subjugation are permissible? Perhaps there are various ways of blocking this inference. For example, in a footnote Nolt acknowledges that one must still have a high level of respect for a non-sentient human body because it is "[. . .] something significant to those who cared about the person of whom it is the remains [. . .]. Similarly, one might owe it respect for what it once was [. . .]" (2006, p. 363). In short, this strategy would require us to assess different moral scenarios on a case by case basis.

A third strategy is to accept that not all autopoietic powers are valuable to organisms in the same way, but to deny that they are of value in varying degrees. Such a position would be coherent providing that the values of various autopoietic powers to various possessors are incommensurable.[23] To help motivate this claim, perhaps one could maintain that because the autopoietic powers of different kinds of organism are so diverse, it would be unrealistic to suppose that they are all valuable to their possessors in a comparable way. Both the second and third strategies would help to accommodate the intuition that many people no doubt have, which is that the autopoietic powers of some very simple living beings, such as a unicellular organism, are surely not of value to them in the same way that survival is of value to, say, a human being.[24]

There is much more to be said about each of these strategies, and as mentioned above, my aim is not to adjudicate them here. Indeed, my aim is not to endorse the inherent value account of organic power but rather to identify it as a hypothesis for future investigation by teleologists. It is a highly revisionary proposal in the sense that it implies that the notion of an organism is not *purely* a biological concept, but in part a metaphysical one since it has evaluative and moral import. But in defence of this feature, it should be noted that Oderberg's notion of flourishing and intrinsic purpose, on which his account of immanent causation turns, is itself steeped in metaphysics. Moreover, if the notion of an organism were in part a metaphysical one, this might explain why biologists and philosophers of biology sometimes do not agree on what falls under the concept of an organism. The explanation would be that the disagreements arise from a difference in underlying metaphysical assumptions.

[23]This is a strategy that Nolt acknowledges briefly in a footnote (2009, p. 146, fn. 25).

[24]As Anne Sophie Meincke has pointed out to me, in the case of cells that are part of a larger organism, the case is much less clear. For example, there is the phenomenon of self-sacrifice of cells in multicellular organisms. Hence, it may be that part-whole relations affect the evaluative status of a cell's own autopoietic powers. Unfortunately, I must postpone a detailed discussion of such cases for another paper.

13.5 Might the Organic/Inorganic Distinction be Fuzzy After All?

In the previous section we considered how one might try to refine the account of immanent causation by tying the notion of flourishing to the idea of autopoietic powers (and perhaps others) being of inherent value to their possessors. Such an account allows us to give a powers-based criterion for organisms, providing that only organic powers are inherently valuable to their possessors. However, we also identified some costs and concerns regarding this theory. It is therefore worth considering the state of play if we are forced to abandon the idea that a sharp distinction can be drawn between organic and inorganic powers. Would this abandonment mean that the notion of an organism is useless or that we must be anti-realists about organisms? Fortunately, I believe the answer is 'no' and that this is best seen by briefly considering John Heil's (2003) theory of higher-level powers. Importantly, such a view would allow us to preserve the idea that the organic/inorganic distinction has something to do with powers of self-maintenance, which following Nolt we have called autopoietic powers. However, the theory would say that the difference between organic and inorganic powers is not sharp. Such a view would also support our earlier suggestions that notions like 'intrinsic purpose' seem to apply in different ways in different cases. Surely plants have purposes in a weaker sense than a human being with conscious intentions.

The discussion of the water cycle also suggested that it is not easy to find a significant difference between the kind of teleology which (we are assuming) is displayed in some inorganic systems and the kind of teleology which is present in organic cases. This again might lead us to think that the difference between various cases of teleology is not a sharp one. Perhaps the ways in which the water cycle maintains itself are simpler and less diverse than the structurally complex self-regulating powers we find in paradigmatic organic cases, which is why we are less inclined to count the water cycle as organic and not inclined to say that the water cycle flourishes in the same way that paradigmatic organisms do. However, if the difference is not sharp, we might expect there to be borderline cases, and this seems to be borne out by the fact that not everyone in biology and philosophy of biology agree on what counts as an organism. For example, historically there has been disagreement between philosophers of biology about whether ecosystems should be regarded as high-level organisms. Clements (1905) was of the first to argue that ecosystems are organisms of sorts, though this proposal has fallen out of favour in recent times. At lower levels there is, for example, the ongoing debate about whether viruses are alive (see Villarreal 2004 for an introduction to this debate).

If we abandon the thought that the distinction between organic and inorganic powers can be clearly delineated, does this mean that the concept of being an organism should be eliminated? I do not think so. To see why, it is instructive to consider Heil's theory of powers in the special sciences. Heil (2003) urges that it is a mistake to think that any special science predicates pick out unique properties which are strictly shared by all the subjects satisfying the predicate. To think otherwise is to be seduced by the "picture theory of language" (2003, pp. 5–7) according to which all predicates pick out distinct unitary properties. According to Heil, the picture theory typically leads to a 'levels' conception of reality, on which each special science concerns a distinct level of nature containing a unique set of autonomous powers. I shall not discuss Heil's reasons for thinking that the levels conception is problematic, nor shall I discuss the claim that the picture theory leads to such a conception. The important question for our purposes concerns what would follow if there is no one property (such as 'having an autopoietic power') which all and only organisms share.

Heil's theory of higher-level predicates is that they do not pick out identical properties that things satisfying the predicate share, but rather they track "similar-but-not-precisely-similar properties" (2003, p. 27). Since Heil is a powers theorist, for him these less-than-perfect-similarities are similarities in respect of the powers of things.[25] To illustrate, Heil explains the higher-level phenomenon of redness as follows: "By virtue of possessing similar-but-not-precisely-similar-properties, red objects possess similar-but-not-precisely-similar 'causal powers' or dispositionalities, and so behave (colourwise) in similar-but-not-precisely-similar ways" (2003, p. 28). What Heil is keen to emphasise, though, is that these similarities are not invented by us but rather are out there in the world: "We do not 'carve up' the world in the sense of manufacturing divisions where none previously existed, but we do commemorate boundaries that, for us, stand out" (2003, p. 49). Importantly, because higher-level predicates track similarities which are objective, Heil insists that he remains a realist about higher-level entities. Higher-level predicates truly apply to the world and so we can still say that there are tables, trees, and so on (Heil 2003, p. 58). The point is just that these higher-level concepts do not mark "hard-edged features of the world" and are to a large degree "vague or non-specific" (2003, p. 58). In other words, the boundaries between high-level kind concepts are fuzzy: "Concepts, and words used to express these, are in most cases satisfied by endless *similar* things; and similarity grades off imperceptibly to dissimilarity" (2003, p. 49).

Although Heil does not discuss the specific property of being an organism, it is not difficult to see what this concept would look like within Heil's framework. The idea would be that organisms share similar-but-less-than-similar powers. If what we have been saying about organic powers earlier is plausible, the powers in question will, first and foremost, be the autopoietic ones. However, since the concept of autopoiesis is itself a high-level one, there will be no one thing that different autopoietic powers strictly have in common. They will each differ in some ways but not in others.[26] Nonetheless, these less-then-precise-similarities will be objective

---

[25]Heil (2003, ch. 11) maintains that powers are identical with qualities, but this detail is unimportant for our purposes.

[26]To use a term famously introduced by Wittgenstein (1953), we might say that the concept of autopoiesis—and hence the concept of an organism—is a family resemblance concept that is not susceptible to strict analysis in terms of necessary and sufficient conditions.

and hence it will be true to say that there are organic powers. Importantly, this framework would allow there to be inorganic cases involving powers which have much in common with those in organic cases (as in the case of sophisticated robots), but not quite similar enough in, say, organisational structure to merit being called organic. We would also expect there to be borderline cases in this framework, in which it is unclear whether a given power is sufficiently similar to those in paradigmatic organic cases to merit being classed as organic.

This is only a sketch of a possible position but it is important to note that the earlier discussion of the water cycle lends some weight to this approach, especially if the notion of organic flourishing cannot be developed in a sharp and informative way. My main criticism of Oderberg was that it is unclear why the notion of flourishing and intrinsic purpose is applicable in all organic cases but no inorganic cases. If this is right, it may suggest that the difference between organic and inorganic systemic teleology is not sharp. Perhaps such a position would leave us with a happy compromise. It would free us from the need to draw a sharp distinction between the organism kind and other kinds of systemic teleological system. At the same time, the Heilean framework would arguably still allow us to be realists about organisms.

The approach outlined will also have implications for questions about value, some of which we discussed in the previous section. According to some theorists, only living creatures have intrinsic value (e.g. Regan 2004; Nozick 1981). However, if there is no sharp boundary between organic and inorganic sytemic teleology, and intrinsic value is grounded in teleology, then the question of what is and is not intrinsically valuable is no longer clear-cut. This would help to motivate Davison's intuition (2012, ch. 4) that there is no clear 'cut-off' between that which is valuable and that which is not, and open up new debates about the intrinsic value of different parts of nature. Although I do not have space to explore this here, I suspect that this picture could have profound effects on ethics and lend weight to the idea that many environmental systems are themselves worthy of respect to some degree, for non-instrumental reasons.

13.6 Conclusions

In this chapter I have explored some possible ways of developing a powers-based theory of organisms and the challenges they face. It would be unrealistic to try to settle the debate here, but we have identified some of the key issues on which the debate is likely to turn. We have seen how the powers theory is highly conducive to a teleological world view of both systemic and non-systemic final causes (one which need not rely on a controversial commitment to Aristotle's substantial forms). A powers-based theory of organisms is likely to turn on the distinction between organic and inorganic systemic teleology, and we examined possible ways to draw this distinction in a sharp way using a specific notion of immanent causation. We then compared the notion of immanent power with the notion of an autopoietic function.

However, the problem is that it is far from clear that immanent causation cannot occur in inorganic realms, as the example of the water cycle illustrates. Oderberg's response is that it is only in the organic cases that the exercise of immanent causation exhibits intrinsic purpose and leads to flourishing. We therefore explored some possible ways of fleshing out the notion of flourishing in an informative and non-circular way, such as the proposal that organisms flourish when their autopoietic powers (or more directly, their manifestations) are of inherent value to them in a morally relevant sense. This theory faces its own challenges, however. In the light of all these problems, we concluded by considering the state of play if powers theorists are, after all, unable to draw a sharp distinction between organic and inorganic powers. We argued that this would not be a disaster, by considering the matter from the perspective of John Heil's metaphysical framework. Even if the distinction between organic and inorganic powers were to some extent fuzzy, this would not mean that it does not track real similarities and differences in the world.

Acknowledgements I would like to thank Anne Sophie Meincke for detailed and invaluable comments on earlier drafts, and for her patience in waiting for the final version. Thanks also go to Alex Carruth, Nancy Cartwright, Chris Cowie, David Faraci, Louise Hanson, Robin Hendry, Simon James, Stephen Mumford, Ben Page and Ben Smith for helpful discussions on this topic.

References

Ariew, Andre. 2007. Teleology. In *The Cambridge Companion to the Philosophy of Biology*, ed. D. L. Hull and M. Ruse, 160–181. New York: Cambridge University Press.

Aristotle. 2008. *Physics*, ed. D. Bostock. Oxford: Oxford University Press.

Armstrong, David M. 1997. *A World of States of Affairs*. Cambridge: Cambridge University Press.

Ayala, Francisco J. 1970. Teleological Explanations in Evolutionary Biology. *Philosophy of Science* 37: 1–15.

Barandiaran, Xabier E., and Alvaro Moreno. 2008. Adaptivity: From Metabolism to Behavior. *Adaptive Behavior* 16: 325–344.

Barham, James. 2012. Normativity, Agency, and Life. *Studies in History and Philosophy of Biological and Biomedical Sciences* 43: 92–103.

Bedau, Mark. 1992. Where's the Good in Teleology? *Philosophy and Phenomenological Research* 52: 781–806.

Bird, Alexander. 2007. *Nature's Metaphysics: Laws and Properties*. Oxford: Oxford University Press.

Cartwright, Nancy. 1989. *Nature's Capacities and their Measurement*. Oxford: Clarendon Press.

Clements, Frederic E. 1905. *Research Methods in Ecology*. Lincoln: University Publ. Co.

Cooper, John M. 1982. Aristotle on Natural Teleology. In *Language and Logos*, ed. M. Schofield and M. Nussbaum, 197–222. Cambridge: Cambridge University Press.

Davison, Scott A. 2012. *On the Intrinsic Value of Everything*. London: Continuum.

Dowe, Phil. 1995. What's Right and What's Wrong with Transference Theories. *Erkenntnis* 42: 363–374.

Ellis, Brian. 2001. *Scientific Essentialism*. Cambridge: Cambridge University Press.

Feser, Edward. 2010. Teleology: A Shopper's Guide. *Philosophia Christi* 12: 142–159.

Garrett, Don. 1999. Teleology in Spinoza and Early Modern Rationalism. In *New Essays on the Rationalists*, ed. R. J. Gennaro and C. Huenemann, 310–333. New York: Oxford University Press.

Heil, John. 2003. *From an Ontological Point of View*. Oxford: Oxford University Press.

Jacobs, Jonathan. 2010. A Powers Theory of Modality: Or, How I Learned to Stop Worrying and Reject Possible Worlds. *Philosophical Studies* 151: 227–248.

Kroll, Nick. 2017. Teleological Dispositions. In *Oxford Studies in Metaphysics, Vol.* 10, ed. K. Bennett and D. W. Zimmerman, 3–37. Oxford: Oxford University Press.

Lennox, James G. 1993. Darwin was a Teleologist. *Biology and Philosophy* 8: 409–421.

Lowe, E. Jonathan. 1980. Sortal Terms and Natural Laws. *American Philosophical Quarterly* 17: 253–260.

———. 1982. Laws, Dispositions and Sortal Logic. *American Philosophical Quarterly* 19: 41–50.

———. 2009. *More Kinds of Being: A Further Study of Individuation, Identity, and the Logic of Sortal Terms*. Oxford: Wiley Blackwell.

Martin, Charles B., and Karl Pfeifer. 1986. Intentionality and the Non-Psychological. *Philosophy and Phenomenological Research* 46: 531–554.

Maturana, Humberto, and Francisco Varela. 1980. *Autopoiesis and Cognition: The Realization of the Living (Boston Studies in the Philosophy of Science* 42*)*. Dordrecht: D. Reidel Publishing Company.

McDowell, John. 1985. Values and Secondary Qualities. In *Morality and Objectivity*, ed. Ted Honderich, 110–129. London: Routledge.

Meincke, Anne Sophie. 2019. *Autopoiesis, Biological Autonomy and the Process View of Life. European Journal for Philosophy of Science* 9: 5. https://doi.org/10.1007/s13194-018-0228-2.

———. This volume. Powers, Persistence and Process. In *Dispositionalism. Perspectives from Metaphysics and the Philosophy of Science (Synthese Library* 417*)*, ed. A. S. Meincke, 89–113. Cham: Springer.

Molnar, George. 2003. *Powers: A Study in Metaphysics*, ed. S. Mumford. Oxford: Oxford University Press.

Mumford, Stephen. 2004. *Laws in Nature*. London: Routledge.

———. 2005. Kinds, Essences, Powers. *Ratio* 18: 420–436.

Mumford, Stephen, and Rani L. Anjum. 2011. *Getting Causes from Powers*. Oxford: Oxford University Press.

Nicholson, Daniel J. 2014. The Return of the Organism as a Fundamental Explanatory Concept in Biology. *Philosophy Compass* 9: 347–359.

Nolan, Daniel. 2015. Noncausal dispositions. *Noûs* 49: 425–439.

Nolt, John. 2006. The Move from *Good* to *Ought* in Environmental Ethics. *Environmental Ethics* 28: 355–374.

———. 2009. The Move from *Is* to *Good* in Environmental Ethics. *Environmental Ethics* 31: 135–154.

Nozick, Robert. 1981. *Philosophical Explanations*. Cambridge: Harvard University Press.

Oderberg, David S. 2007. *Real Essentialism*. London: Routledge.

———. 2008. Teleology: Inorganic and Organic. In *Contemporary Perspectives on Natural Law*, ed. A. M. González, 259–279. Aldershot: Ashgate.

———. 2010. The Metaphysical Foundations of Natural Law. In *Natural Moral Law in Contemporary Society*, ed. H. Zaborowski, 44–75. Washington, DC: Catholic University of America Press.

———. 2017. Finality Revived: Powers and Intentionality. *Synthese* 194: 2387–2425.

Regan, Tom. 2004. *The Case for Animal Rights*. Berkeley: University of California Press.

Ruse, Michael. 1989. Do Organisms Exist? *American Zoologist* 29: 1061–1066.

Singer, Peter. 1990. *Animal Liberation*. 2nd ed. New York: New York Review of Books.

Taylor, P. W. 1986. *Respect for Nature: A Theory of Environmental Ethics*. Princeton: Princeton University Press.

Vetter, Barbara. 2015. *Potentiality: From Dispositions to Modality*. Oxford: Oxford University Press.

Villarreal, Luis P. 2004. Are Viruses Alive? *Scientific American* 291: 100–105.

Walsh, Denis. 2008. Teleology. In *The Oxford Handbook of Philosophy of Biology*, ed. M. Ruse, 113–137. New York: Oxford University Press.

Williams, Neil E. This volume. What Are Manifestations? In *Dispositionalism. Perspectives from Metaphysics and the Philosophy of Science (Synthese Library* 417*)*, ed. A. S. Meincke, 67–87. Cham: Springer.

Wittgenstein, Ludwig. 1953. *Philosophical Investigations*, ed. G. E. M. Anscombe and R. Rhees, trans. G. E. M. Anscombe. Oxford: Blackwell.

Author Index

A. S. Meincke (ed.), *Dispositionalism*, Synthese Library 417,
https://doi.org/10.1007/978-3-030-28722-1

Subject Index

A. S. Meincke (ed.), *Dispositionalism*, Synthese Library 417,
https://doi.org/10.1007/978-3-030-28722-1

The manufacturer's authorised representative in the EU is Springer Nature Customer Service Centre GmbH, Europaplatz 3, 69115 Heidelberg, Germany. If you have any concerns regarding our products, please contact ProductSafety@springernature.com

Printed and bound by CPI Group (UK) Ltd, Croydon, CR0 4YY
30/09/2026
02225445-0006